18.50 + TAX

John Burroughs at Slabsides, the cabin retreat he built in the woods near his home in West Park, New York

A SHARP LOOKOUT

SELECTED NATURE ESSAYS
OF
JOHN BURROUGHS

EDITED BY FRANK BERGON

SMITHSONIAN INSTITUTION PRESS

WASHINGTON, D.C.

LONDON

Library of Congress Cataloging-in-Publication Data

Burroughs, John, 1837–1921.
  A sharp lookout.

  Bibliography: p.
  1. Natural history. I. Bergon, Frank. II. Title.
QH81.B966 1987 508   87-43076
ISBN 0-87474-270-6
ISBN 0-87474-271-4 (pbk.)

British Library Cataloguing-in-Publication Data is available.

The paper used in this publication meets the minimum
requirements of the American National Standard
for Permanence of Paper for Printed Library Materials
Z39.48–1984.

Photographs in this book are reproduced through the
courtesy of Elizabeth Burroughs Kelley, James Stapleton,
and the Vassar Archives.

CONTENTS

To Elizabeth Burroughs Kelley

A NOTE ON THE TEXT

The text of the essays in this volume is reprinted
from the Riverby Edition of *The Writings of John
Burroughs* published by the Houghton Mifflin Company in twenty-three volumes from 1904 to 1922.
Burroughs frequently corrected and revised his
essays between their printings in periodicals and in
books, and the uniform Riverby edition represents
the essays in their final format. Other texts as well
as sources for directly quoted passages in the
Introduction and the Editor's Notes are documented
in the Bibliographical Note at the back of this book.

F.B.

On the porch of Slabsides with a favorite pet

INTRODUCTION

FRANK BERGON

John Burroughs wrote no sustained masterpiece, no *Walden*, no *Selborne*, but during a writing career of over sixty years he achieved recognition as the dean of American nature writers. He published nearly three hundred essays that eventually helped fill twenty-seven books, and despite his caution to his biographer ("Remember my books are not the corner-stone of American literature") his finest essays stand with those of Thoreau and Muir as the best examples of the nature essay written in the nineteenth century, a period when the genre blossomed and numerous authors contributed to its popularity. Each reader may have a favorite Burroughs book—*Pepacton, Signs and Seasons,* and *Locusts and Wild Honey* contain some of his best essays—but no single volume of essays emerges as an undisputed classic. Even in individual essays, Burroughs is most effective in short bursts. The purpose of this collection is to gather for the first time some of Burroughs's finest nature essays, while simultaneously doing justice to the range and variety of his long career as a nature writer.

Before his death in 1921 Burroughs had achieved considerable fame as the preeminent naturalist of the Catskill Mountains and the Mid-Hudson Valley. Special editions of his essays became required reading in schools, and Burroughs Nature Clubs multiplied across the country. People made pilgrimages to visit Burroughs at Riverby, the home he had designed and helped build at West Park overlooking the Hudson, or at Slabsides, the cabin retreat he built in the woods west of his home. Many also went to Woodchuck Lodge, the country house he renovated on his boyhood farm in the Catskills. Walt Whitman came as a friend in the 1870s, and Theodore Dreiser came as a reporter in the 1890s. In 1903, twenty-one years after Oscar Wilde visited Burroughs at Riverby, a government yacht brought President Theodore Roosevelt up the Hudson for a day's visit. The guest books at Slabsides alone record the signatures of nearly seven thousand visitors. On a single spring day in 1915 there arrived, as Burroughs noted in his journal, "One hundred and ten Vassar girls yesterday at Slabsides, and a dozen High School girls from Poughkeepsie, also President MacCracken [of Vas-

INTRODUCTION

sar] and his wife," all traveling across the river from Poughkeepsie to visit John O'Birds, as he had come to be called.

It is difficult to exaggerate the extent of Burroughs's fame. In the late 1930s, Fred Eastman published the biographies of twenty men in a five-volume set entitled *Men of Power*. The fourth volume included biographies of Lincoln, Tolstoy, and John Burroughs, the man who "had learned what was essential and inessential to the art of living. . . . Men and women who had been battered and bruised in the noisy scramble of American cities heard in Burroughs a call to the simple life. Here was a man who had kept his soul. They would seek him out." What many people sought to find in Burroughs were the qualities they attributed to nature, which was "a beneficent mother," as Eastman said, "to be cherished."

Those qualities were further defined by William Dean Howells, who found Burroughs's first book of nature essays, *Wake-Robin*, "fresh, wholesome, sweet, full of gentle and thoughtful spirit." If many people felt displaced by a scientific view of nature and by evolutionary theory, John Burroughs was the man, as Henry James wrote in his review of *Winter Sunshine*, "who sees sermons in stones and good in everything." Little wonder that Burroughs

11

the celebrity could be imagined, in Whitman's phrase, as "stucco'd with quadrupeds and birds all over."

"My wife is always pointing out the bird's nest in my hair," Burroughs wrote to a friend, "but she doesn't see deep enough—it is in my brain." Similarly, the popular image of the Sage of Slabsides does not sufficiently reveal the deeper, more interesting figure who emerges from the pages of Burroughs's writings. In contrast to those who wanted only sermons from stones, Burroughs indirectly defined his goals as a writer while praising Wordsworth, whose "pages are full of real objects and their relations, not mere fancies or sentiments or vague longings of vapoury refinements, but solid reality."

It was solid reality Burroughs wanted in his own work, and people who had read few or none of his essays, yet swooned over his youthful poem "Waiting," caused him to say, "I almost wish I had never written the poem—they pester me so about it." On learning that a woman in Poughkeepsie died with the verses in her hand, Burroughs confided to his journal: "Pure poetry never affects people in this way. . . . My little poem is vague enough to escape the reason." What Burroughs repeatedly stresses in his journals and essays is that the starting point

INTRODUCTION

of his work is the concrete, the real, "the great, shaggy barbarian earth" itself. "I do not look for sermons in stones," he wrote. "I paint the bird for its own sake, and for the pleasure it affords me, and am annoyed by any lesson or moral twist." Although readers can find lessons in Burroughs's pages, they are not always the expected ones. "The universe," Burroughs tells his readers, "is no more a temple than it is a brothel or a library." It is not a symbol of anything. Nature is the primary and everlasting fact, and we, as living beings and a species, are secondary and temporary facts, "mere ephemera of an hour, like the insects of a summer day." It was such a view that allowed Burroughs to write in an early essay that nature doesn't care a fig for any creature over another, and man is no exception; and it allowed him to write in one of his last books, *Accepting the Universe,* "I love Nature, but Nature does not love me."

Writing about Burroughs's early book *Wake-Robin* (1871), the ornithologist and historian Elliott Coues came closest to appreciating Burroughs's achievement: "I . . . can bear witness to the minute fidelity and vividness of your portraiture. How many things you saw—how many more you *felt.* . . . You bring it all back to me—things which I felt at the time, but which passed like last night's

dream." Burroughs offered his book to the public
as simply a careful and conscientious record of
actual observations and experiences, and he was
pleased that the book gave the noted ornithologist
the experience and "rare pleasure of enjoying the
birds as nowhere else excepting in the woods and
fields—where you carry me straightway." Coues's
observations underscore the distinction Burroughs
frequently made between the mere thinker and the
artist. The thinker "tells" us about a thing, Burroughs says, while the artist "gives" us the thing.
It was as such an artist that Burroughs defined
himself, and it is as such that he should be judged.
He was a writer, a man of letters whose work
includes numerous essays on science, travel, religion, literature, and philosophy, but whose central
subject was the concrete natural world and his own
experience of that world.

John Burroughs was born in 1837, the seventh
of ten children, on his family's three-hundred-acre
dairy farm near Roxbury, New York. As a boy,
Burroughs knew he wanted to be a writer, although
he had little idea of what kind of writer he might
be. His father was an old-school Baptist whose
reading consisted of the newspaper, the Bible, a
hymnal, and a monthly religious paper, *The Spirit
of the Times*. Books were of little importance to the

INTRODUCTION

family, whose ancestors of English, Irish, and Welsh stock had been primarily farmers in New England and New York since the seventeenth century. Even after Burroughs became famous, neither his parents nor his brothers and sisters read any of his books. Burroughs later considered his life as a farm boy, his rural chores and daily encounter with nature, the most important experience of his life, and the Old Home, as he called his family's farm, formed the emotional center of his later life. As a farm boy, he grew up accepting man and his tools as an intrinsic part of the natural world. His portrayals of men mowing hay or plowing fields are presented as rural pageants, examples of men moving to the rhythms of nature. "Take that farm-boy out of my books," he said, "and you have robbed them of something vital and fundamental."

After periodically attending district schools, Burroughs left home at seventeen to become a teacher in a country school in Ulster County. For the better part of the next nine years he taught in various country and small-town schools, mostly in New York, but also briefly in Illinois and New Jersey. With savings from his teaching, he attended Hedding Literary Institute in Ashland, New York, for three months and Cooperstown Seminary for the same length of time. This schooling was the extent of his directed study except for a few months he spent studying medicine with a physician while continuing to teach at a nearby school.

Unlike many incipient writers, Burroughs was immediately attracted to the essay as a literary genre, and his first efforts, published in small newspapers, showed the influence of Samuel Johnson and, later, of Addison and Lamb. When he was nineteen, he discovered Emerson. He had read Emerson before, but this time he "read him in a sort of ecstasy. I got him in my blood and he colored my whole intellectual outlook." More practically, Emerson's example helped Burroughs move away from an imitation of Johnson's periodic sentences to a plainer, more direct style, and helped confirm Burroughs in his literary ambitions. In 1857, shortly after his marriage, the twenty-year-old Burroughs confessed in a letter to his wife: "I sometimes think I will not make the kind of husband that will always suit you. If I live, I shall be an author. My life will be one of study. . . . I know I must struggle hard to realize my end, to have my name recorded with the great and good. But if God spares my life, the great world shall know that I am in it."

Recognition from the great world almost came in 1860, when he published his first essay in a national magazine. But as Burroughs reports, the essay,

INTRODUCTION

called "Expression," was so Emersonian that James
Russell Lowell, the editor of *The Atlantic Monthly*,
held up publication to check through Emerson's
writings for possible plagiarism. The essay was
published without a byline, and to this day it is
attributed to Emerson in Poole's *Index*. Burroughs
received advice from an older writer, David Was-
son, to put aside the writing of reflective essays
until his maturer years; and this advice, coupled
with his own observation that the imitator is doomed
to mediocrity, led to his first essays about rural life:
"It was mainly to break the spell of Emerson's
influence and to get on ground of my own that I
took to writing upon outdoor themes." His series
of essays called "From the Back Country," which
concerned stone walls, the making of maple sugar
and butter, and other aspects of farm and country
life, began appearing in the New York *Leader*. But
the spell of Emerson remained evident in reflective
essays that Burroughs continued to publish in the
New York *Saturday Press* and *Knickerbocker Magazine*.

Even after he found his true mode of expression
in the nature essay, Emersonian analogies and
epigrams occasionally appear in his work, as in
"Touches of Nature," but Burroughs's method and
his view of nature are usually the opposite of

Emerson's. Their views also differ concerning the nature of man as defined by man's relationship to nature. In writing about the "natural history of the boy" in "Touches of Nature," Burroughs depicts the boy as part of nature, rather than presenting nature as part of the boy. We don't read nature, he frequently noted; nature reads us. Burroughs felt that both Emerson's distance from physical nature and his passion for analogy frequently betrayed him into seeing the whole of nature as only a metaphor for the human mind. Despite his own strain of idealism, Burroughs did not accept such a dualistic view of the universe, especially a view that made nature auxiliary to man. Yet he never lost his admiration for Emerson, and all his life acknowledged his debt to his early mentor. Three of Burroughs's published essays (and portions of several others) were about Emerson, and his last writing was the concluding paragraph of an essay about Emerson's *Journals*.

Burroughs met Emerson for the first time in 1863, after a lecture by Emerson at West Point; but that year is significant in Burroughs's career for other reasons. In the same year, while teaching at nearby Highland Falls, Burroughs discovered John James Audubon's *Birds of America* in the library at West Point. The discovery, as Burroughs said, was "like bringing together fire and powder." Audubon's paintings and writings ignited Burroughs's interest in ornithology. Burroughs thought that verbosity and affectation occasionally marred Audubon's prose, but his enthusiasm, similar to Burroughs's own youthful reaction to birds and other wildlife, made ornithology an adventure.

At this time Burroughs's letters also show a growing interest in botany, and he had begun his first efforts in the writing of natural history before he moved to Washington, D.C., where after a brief job with the Quartermaster General's Department, he began work for the Currency Bureau of the Treasury Department. For most of the next nine years, until he returned to settle permanently in the Mid-Hudson Valley, Burroughs's work as a clerk and later a bureau chief placed him in front of a steel vault guarding millions of dollars in government bank notes. The most significant break from the routine came when he was assigned to transfer government bonds to England and thus made his first trip abroad. Otherwise, he guarded the vault. "How I reacted against the door of that old safe," he said. "But the rebound sent me back to the fields and woods of my boyhood." The job gave him time to write, and as with many writers, distance from his material seemed to help Burroughs

INTRODUCTION

find his true subject. He met Walt Whitman in
1863, and also this friendship solidly fused Burroughs's interest in natural history with his literary
ambitions.

Burroughs and Whitman became both mutual
friends and literary advisors. Whitman accompanied Burroughs on walks in the woods, and
Burroughs accompanied Whitman on visits to soldiers wounded in the Civil War. Burroughs's description of the song of the hermit thrust provided
Whitman with a major figure and symbol in "When
Lilacs Last in the Dooryard Bloom'd," and Whitman encouraged Burroughs to entitle his first book
of nature essays after a wildflower, the wake-robin,
whose appearance in the spring coincides with the
arrival of migrating birds in May. More important,
Burroughs became a major promoter of Whitman's
poetry. His 1866 review of *Drum-Taps* was the first
intelligent critical essay to be published on Whitman
in the nineteenth century. The essay still stands as
perceptive criticism, and it was the basis for Burroughs's earliest book, *Notes on Walt Whitman as
Poet and Person* (1867), which Whitman himself
silently helped edit and expand. Subsequently,
Burroughs would show Whitman the articles he
wrote about him prior to publication, a habit certainly not conducive to unbiased or original criti-

cism, but one aligned to Burroughs's purpose of bringing as strong a case as possible before a recalcitrant public.

Whitman later visited Burroughs three times in the Mid-Hudson Valley, and in *Specimen Days,* while describing one of the visits, Whitman raved about the "place, the perfect June days and nights, (leaning toward crisp and cool,) the hospitality of J. and Mrs. B., the fruit (especially my favorite dish, currants and raspberries, mixed, sugar'd, fresh and ripe from the bushes—I pick 'em myself)—the room I occupy at night, the perfect bed, the window giving an ample view of the Hudson and the opposite shores, so wonderful toward sunset, and the rolling music of the R.R. trains, far over there—the peaceful rest." In the mornings, Burroughs and Whitman swam in the Hudson before breakfast, and during Whitman's visit in 1879, Burroughs took him across the river to Vassar to meet Frederic Ritter, a composer and teacher of music, who had set some of Whitman's poems to music. Whitman wrote about his nature walks and countryside drives with Burroughs, remarking that the "grandest sight" was an eagle soaring above the Hudson.

After Whitman's death, Burroughs published another book, *Whitman: A Study* (1896), and continued to write about Whitman in no fewer than

INTRODUCTION

two dozen additional essays. Although Burroughs was the first critic to point out the importance of previous literature to Whitman's development (something the poet was reluctant to acknowledge) and although he observed in his journal the extent to which Whitman worked over his lines, Burroughs noted that the effect of *Leaves of Grass* was seemingly a rejection of all literary artifice and a spontaneous expression from nature itself. He saw Whitman as the great interpreter of nature, and he resolved also to become its interpreter in his own way. In his journal he records a significant conversation with Whitman about the writing of natural history:

I was trying to express to him how, by some wonderful indirection, I was helped by my knowledge of the birds, the animals—cows, and common objects. He could see, he said. The ancients had an axiom that he who knows one truth knows all truths. There are so many ways by which Nature may be come at; so many sides to her, whether by bird, or insect, or flower, by hunting, or science—when one thing is really known, you can no longer be deceived; you possess a key, a standard; you effect an entrance, and everything else links on and follows. He thought natural history, to be true to life, must be inspired as well as poetry.

Burroughs's own interpretation of Whitman's advice becomes clearer in later journal entries where he further equates the writing of natural history with the making of literature: "In one sense the great poet, and the great naturalist, are the same—things take definite and distinct shape to them; they are capable of vivid impressions."

Burroughs's first nature essay, "With the Birds" (revised and reprinted as "The Return of the Birds") appeared in the *Atlantic* in 1865, three years after the death of Thoreau, who invented the form of the essay that Burroughs was to make his own. Although Burroughs's reaction to Thoreau was always mixed, his debt to Thoreau was deeper than he was willing to acknowledge early in his career. He began reading Thoreau in late 1862 or early 1863, before moving to Washington, during that crucial period in his development when he was groping his way toward the nature essay. At that time he became friends with Myron Benton, a writer and farmer in Amenia, New York, whose extensive correspondence with Burroughs continued for roughly the next forty years. Benton had received the last letter Thoreau wrote before he died, and it was Benton who through letters, conversations, and an Adirondack hunting trip with Burroughs in the summer of 1863, encouraged him in his reading of Thoreau and his writing about nature.

INTRODUCTION

Burroughs's early reluctance to admit Thoreau's influence ("I am not conscious of any great debt to Thoreau") is understandable because of the comparison immediately made between the two writers when Burroughs began publishing his nature essays. In a review of *Winter Sunshine,* after praising Burroughs for "originality," a "real savour" of his own, a "real genius for the observation of natural things," and a "style sometimes idiomatic and unfinished to a fault, but capable of remarkable felicity and vividness," Henry James sums up his response by saying, "Mr. Burroughs is a sort of reduced, but also more humorous, more available, and more sociable Thoreau." In a letter to Benton a few years later, Burroughs wrote, "That Thoreau business, I think will play out pretty soon. There is little or no resemblance between us." The main distinction Burroughs always made was that although Thoreau "prized the fact, and his books abound in delightful natural history," Thoreau was a born supernaturalist rather than a born naturalist; he was always looking "intently for a bird behind the bird,—for a mythology to shine through his ornithology."

It may be fairer to Thoreau to say he was interested in both bird *and* a bird behind the bird, while Burroughs was more intently interested in

INTRODUCTION

the bird itself and his experience of it. There is a passage in *Walden* where Thoreau describes a feeling of "savage delight" when he caught a glimpse of a woodchuck and was "strongly tempted to seize and devour him raw; not that I was hungry then, except for the wildness which he represented." In "Birch Browsings," Burroughs, no doubt establishing his difference from Thoreau, describes a moment when he, too, saw a woodchuck but experienced hunger for something other than the animal's symbolic value. "I slaughtered him just as a savage would have done and from the same motive,—I wanted his carcass to eat."

Burroughs's own essays on Thoreau, despite their fault-finding stance, mark perhaps the best critical assessment Thoreau received in the United States before the twentieth century. Burroughs stressed a consideration of Thoreau's writings as literature. He was alert to Thoreau's humor and valued his modes of "tall talk," extolling *Walden* as "certainly the most delicious piece of brag in our literature." He praised Thoreau's style as even more expressive than Emerson's, although he thought Thoreau's extreme uses of exaggeration and paradox were merely words devoid of meaning. He emphasizd that Thoreau gave us not a philosophy but a life, one that was fervent and honest.

INTRODUCTION

But when we regard Thoreau simply as an observer or as a natural historian, there have been better, Burroughs says, "though few so industrious and persistent." Thoreau kept nature around Concord "under a sort of police surveillance." He "watched the approach of spring as a doctor watches the development of a critical case." The results, however, were frequently disappointing, especially when Thoreau's ear was at work rather than his eye. Burroughs's list of Thoreau's errors in natural history is long, and yet, even though Thoreau confused the song of the hermit thrush with that of the wood thrush, and although any schoolboy could have told him that the song of his phantom night warbler was that of the oven bird, Thoreau "could describe bird-songs and animal behavior and give these things their right emphasis in the life of the landscape as no other New England writer has done." In *Cape Cod* and *The Maine Woods,* according to Burroughs, Thoreau gave us nature writing at its best, and in *Walden* he "gave us our first and probably only nature classic."

When he was eighty-three, Burroughs gave his biographer, Dr. Clara Barrus, a list of writers to whom he was indebted, "those who have helped me to find myself." The list included Thoreau. While always maintaining his difference from Tho-

INTRODUCTION

reau, Burroughs was disturbed by anyone who praised his own work at the expense of Thoreau's. In comparing his own "Exhilarations of the Road" with Thoreau's "Walking," he judged Thoreau's essay to be "infinitely better," pointing out that Thoreau walks into all sorts of fields, into the land of mythology, ideality, religion, "while I just walk in the fields and woods." At another time Burroughs said, "I am like a mellow, sweetish, pleasantly-flavored apple beside the crisp, tart, snappy Spitzenberg, which requires good teeth and crackles as you bite"; but Burroughs added, "I think it probable my books send people to nature more than Thoreau's do."

When Burroughs began publishing his books, the nature essay was accepted as a polite literary genre, practiced in varying degrees and with varying success by writers such as James Elliott Cabot, Thomas Wentworth Higginson, Wilson Flagg, James Russell Lowell, Gail Hamilton, Charles Dudley Warner, and Donald Grant Mitchell. Compared to Thoreau, Burroughs found other nature writers "tame and insipid." He objected to the "dainty euphemisms" and "sentimental gallantries" that marred much nature writing. He thought that James Russell Lowell wrote more about the nature he saw in books than in the outdoors. Higginson made his

INTRODUCTION

birds sit prettily on limbs like stuffed specimens. Nature itself often disappeared behind a veneer of moral and religious reflections. What such nature writing lacked were the qualities Burroughs found in the best literature: "perfect seriousness and singleness of purpose."

Burroughs's stated purpose in his first book of nature essays is "to present a live bird,—a bird in the woods or the fields,—with the atmosphere and associations of the place, and not merely a stuffed and labeled specimen." What he immediately brought to the nature essay was a fresh and intimate knowledge about the natural world. This knowledge of facts and a respect for accuracy allowed him to present nature as more than a storehouse of moral and intellectual truths or pretty scenery. Burroughs found a model for such writing in *The Natural History of Selborne* by Gilbert White, who valued his facts, as Burroughs says, for what they were, not for any double meaning he could wring out of them, or any airy structure he could build upon them. "He loved the bird, or the animal, or a walk in the fields, directly and for its own sake." To those in search of nature's religious or ethical significance, Burroughs would be inclined to reply that it means itself. His own reflections on the natural world were directed less toward discovering moral significance than toward discovering the natural reasons for events and habits.

The representation of charming and picturesque scenes that defined much nature writing was also secondary to Burroughs. "Scenery hunting is the least satisfying pursuit I have ever engaged in," he wrote, and although he occasionally comments on picturesque and sublime elements in nature, he presents them as nature's indirect surprises. The scenic beauty of the Catskills and the Hudson River had been firmly established by Thomas Cole, Asher Durand, Frederick Church, and other artists associated with the Hudson River School of painting, but Burroughs emphasized another aspect of nature not always perceived by the admirers of these painters. "Beauty without a rank material basis enfeebles." Burroughs notes that in Greek mythology Beauty is represented as riding on the back of a lion. "Before beauty is power," he writes, saying it is a mistake to treat beauty as an end in itself or to separate it from what is vital and dramatic in nature.

Nature as drama, then, became the focus of Burroughs's essays, and his own effort to dramatize nature replaced much of the moralizing of other writers. In one of his early "From the Back Country" sketches, while writing about the tame proceedings of country life, Burroughs expressed a desire for a dramatic encounter with a mountain panther or with an Indian. In later essays he shows that nature is full of "little dramas and tragedies and comedies . . . if our eyes are sharp enough to see them." He speaks of certain birds as "actors behind the scenes" and writes about "tragedies of the nests," but he emphasizes that such dramatization must not distort or falsify the facts of natural history. A recognition of migratory birds as "struggling, living beings" allows him to see that it is not untruthful to view their lives as a "series of adventures and hair-breadth escapes by flood and field."

A sense of the dramatic often informs what might be presented by other writers as a static scene. The breaking up of ice on the river, the opening of flowers, and the arrival of spring birds become to Burroughs a "great open-air panorama." At times Burroughs remains an observer of this drama. At other times he becomes a participant, an actor himself, as when he confronts a black snake in "The Return of the Birds," or takes his solitary voyage in a homemade boat down the east branch of the Delaware River in "Pepacton." His essays are full of "bursting, growing things," and, in "A

INTRODUCTION

Bunch of Herbs," he is even able to perceive a dramatic adventure in the lives of the most common weeds.

"Vital" is one of Burroughs's favorite words, and he applies it to everything he values most deeply. "Exact facts are good," he says, "but vital facts are better." Of all wild creatures, birds seem of special importance to him because they are examples of life at the highest pressure. "A bird lives so intensely . . . large-brained, large-lunged, hot, ecstatic, his frame charged with buoyancy and his heart with song . . . a bird seems at the top of the scale." In Burroughs's scheme of values the opposite of vitality is not necessarily death or the inanimate. A tree killed by lightning bears witness to nature's vitality, as do the struggles and deaths of animals—all instances of nature as becoming, as he says in an early essay, where there is no pause, no completion, no explanation.

What repels Burroughs is stagnancy and decay. The vastness of the sea and the darkness of Mammoth Cave may frighten him and incline him to thoughts of death, but the void of the sea is reminiscent of moments before creation, without decay, and Mammoth Cave is a dramatic product of elemental forces. On the other hand, some of the desolate landscape of the West strikes him as inert refuse, and the perpetual summer of Jamaica's jungles oppresses him as a form of stagnation. He prefers the vital days of April to the cloying ones of August. He describes in "August Days" how the algae on the Hudson are at their rankest, and in the woods the decay of mushrooms becomes a "kind of unholy rotting ending in blackness and stench." He describes a woodchuck in terms appropriate to a kind of fungus: "such a flaccid, fluid, pouchy carcass I have never before seen . . . baggy and shaky as a skin filled with water. Let the rifleman shoot one while it lies basking on a sideling rock, and its body slumps off, and rolls and spills down the hill, as if it were a mass of bowels only." August days are when woodchucks are "most fat and lazy." Winter days, on the other hand, make the country "more of a wilderness, more of a wild solitude," but "warm jets of life still shoot and play amid this snowy desolation." Foxes, hares, skunks, partridges, squirrels, mice, crows, grosbeaks, jays, and snow buntings abound, and even in the frozen river, "when the nights are coldest, the ice grows as fast as corn in July."

Burroughs always described himself as an "essay-naturalist" or a "literary naturalist" to distinguish his work from that of the strict scientist or naturalist. But in the "great writer, in whatever field," Bur-

INTRODUCTION

roughs says, "we encounter real things, real values, real differences, real emotions, real impressions; his sense of reality always saves him from phantoms." Burroughs notes that the greater the writer, the more accurate are the details of the work. Goethe was more correct in his specifications of nature than was Longfellow, and, as Burroughs attempted to demonstrate through numerous examples, Shakespeare was the most knowledgeable and accurate of all in his incidental allusions to nature. Accuracy of specification, however, is not enough to make literature. Only the writer, says Burroughs, who looks upon the real with passion, with emotion, will transmute it into literature, for in the end it is not the subject that makes literature, but the author. Common facts and experiences are changed and heightened according to the writer's quality of mind, and just as there is no exact science "without the clear, white light of understanding," so too "there is no live literature without the play of personality."

The same holds true for the literary naturalist. To make literature out of natural history, accuracy of observation and representation is the main requisite; but since literature does not grow wild in the woods, every artist does more than copy nature. He does not take liberties with facts—for these are the flora upon which he lives—but Burroughs says, "I must give them my own flavor. I must impart to them a quality which heightens and intensifies them." The analogy he makes most frequently is between the artist and the bee. "The hive bee does not get honey from the flowers; honey is the product of the bee. What she gets from the flowers is mainly sweet water or nectar; this she puts through a process of her own, and to it adds a minute drop of her own secretion, formic acid. It is her special personal contribution that converts nectar to honey. . . . Her product always reflects her environment, and it reflects something her environment knows not of. We taste the clover, the thyme, the linden, the sumac, and we also taste something that has its source in none of these flowers."

Although not original in his theories, Burroughs was a more conscious and sophisticated literary craftsman than is usually acknowledged. The apparent artlessness of his writing—the casual offerings of a naturalist who takes the reader for a walk in the woods—disguises the actual skillfulness of his descriptions of the natural world. He tells the reader not to forget the "illusions of all art." And "if my reader thinks he does not get from Nature what I get from her, let me remind him that he can hardly know what he has got till he defines it to himself as I do." For experience to become real it must pass through the imagination and the process of writing itself.

Burroughs's view of good style as stemming from the direct relationship of the writer to language characterizes his own relationship to nature and the relationship he wishes between his books and readers—all three must be "vital, intimate, personal." Burroughs carries out this literary program with as much artful discretion as he carries out his program as a naturalist. "In nature," he writes, "we live and move and have our being. Our life depends upon the purity, the closeness, the vitality of the connection. . . . Artifice, the more artifice there is thrust between us and Nature, the more appliances, conductors, fenders, the less freely her virtue passes." Similarly, it was the literary artifice of much nature writing of his time that he wanted to eliminate from his own work; "language in the hands of the master is transparent."

Such transparency of style and naturalness of expression are themselves the products of conscious literary manipulation. Burroughs stresses that verbs and concrete nouns are the vital elements of language, that adjectives too often get in the way of forceful expression. The archaic, the quaint, the eccentric, the affected are forms of expression he lists as drawing attention only to language itself. Today Burroughs remains readable as much for what he avoids in his prose as for what he adopts. His avoidance of the tangled syntax of Wilson Flagg and the literary diction of T. W. Higginson was a means of trying to invest the nature essay with what he defined as the "basic qualities of good literature—directness, veracity, vitality, and beauty and reality of natural things."

These virtues are not everywhere evident in Burroughs's work, especially in fulsome essays where he is trying to duplicate Whitman. In the extravagant description of "Strawberries," numerous exclamation points only emphasize the lack of genuine emotion issuing from the words themselves. Even his conscious use of the pathetic fallacy does not always rescue his portraits of birds as "widowed mothers" or "happy bridegrooms," although humor and a Darwinian acceptance of nature's amorality do reduce sentimentality. Burroughs's major weakness, though, is often in the organization of his essays. It was a fault he saw in Emerson, whose chief strength, according to Burroughs, was in the making of individual sentences—"electric sparks, the sudden, unexpected epithet or tense, audacious phrase, that gives the mind a wholesome shock,"

INTRODUCTION

but whose argument, as Burroughs says, borrowing a phrase, was a "rope of sand."

In some of Burroughs's essays that begin strongly, the focus and energy dissipate long before the essay reaches a conclusion. In "The Fox," the interesting observation about the fox's ability to thrive under conditions of extermination is not sufficiently investigated, although it is picked up and developed in a later essay. In essays like "Birch Browsings," however, the conclusion is dramatically (and humorously) supported by the fairy-tale quality of actual events in the wilderness. Some essays achieve a coherence and focus by following the rhythm of a single season, as in "The Return of the Birds," or by restricting the dramatic movement of the essay to a single event, like a fishing trip or a journey down a river, as in "Speckled Trout" or "Pepacton." At times even Burroughs's most formless essays work to his advantage, distinguishing him from other writers and freeing him from imposing a point or moral on what he sees as he moves from observation to observation. The experience for the reader of such essays is similar to one announced in the opening sentence of a late essay: "Let us go and walk in the woods."

With this invitation—*let us go and walk in the*

INTRODUCTION

woods—Burroughs shows us that revelations and surprises are near at hand, close to home, just as they were when he was a boy on his family farm in the Catskills, where nature first showed him wonders. Years after the passenger pigeon became extinct, he recalled the days of his boyhood in late March or early April when the "very air at times seemed suddenly to turn to pigeons" and the "naked beechwoods would suddenly become blue with them." Give a child the name of a bird, it has been said, and the child loses the bird. In Washington, D.C., Burroughs felt the loss of his "gladdest hours" as a farm boy when he had first experienced nature without reflection or self-consciousness. "This is my disease," he wrote of his nostalgia for the past, "it is in my system." His longing for his Old Home became a "homesickness which home cannot cure" because it was a longing not just for an extinct bird or a family or a farm, but for a lost boy as well.

In 1873, after Burroughs had left Washington, D.C., he bought a nine-acre fruit farm on the west bank of the Hudson, eighty miles north of New York City. One of the attractions of his new home was that it was only a day's journey from his Old Home in Roxbury, New York. He remained a government employee until 1886, first as a receiver for an insolvent bank in Middletown, not far from his home, and then as a bank examiner, mainly for districts along the Hudson—a job that took up only four or five months of each year. The rest of the time was devoted to farming (growing mostly fancy table grapes) and to writing. He built a bark-covered study near the stone house he called Riverby, and in late 1895, on newly acquired property in a wooded glade a mile and a half from Riverby, he began building a cabin covered with slabs of hemlock and chestnut that would become the guest house and retreat he called Slabsides. During this time Burroughs frequently visited the Old Home, which stayed in the family with his financial help. After 1910 he spent part of each year at the old farm, living in the house he called Woodchuck Lodge and working in his hay-barn study. In 1913, Burroughs's financial worries about his boyhood farm finally disappeared after a gift from Henry Ford made it possible for him to buy the farm outright.

Today, the Riverby bark-covered study, Slab-sides, and Woodchuck Lodge are all designated National Historic Landmarks. Riverby itself and another house on the Riverby property remain the homes of Burroughs's granddaughters, Ursula Burroughs Love and Elizabeth Burroughs Kelley. Slab-

INTRODUCTION

sides has open-house days twice a year, in the spring and fall, on the third Saturday in May and the first Saturday in October. Surrounding Slabsides are one hundred and eighty acres designated as the John Burroughs Nature Sanctuary, open to the public, and managed by the John Burroughs Association. Since 1977, James Stapleton, a professor of environmental science, has lived year-round at the sanctuary and served as its resident naturalist and director of studies. In most ways, the area remains much as Burroughs found it, and even the muck swamp next to the cabin where he grew celery and other vegetables has gone back to brush and trees.

Burroughs called the area around Slabsides "Whitman Land," because "in many ways," he explained, "it is typical of my poet . . . elemental ruggedness, savageness, and grandeur, combined with wonderful tenderness, modernness, and geniality." One of the first guests to stay at Slabsides in 1896 was John Muir, whose talk kept Burroughs up much of the night. Burroughs later said that "you must not be in a hurry, or have any pressing duty" when Muir starts talking. "Ask him to tell you his famous dog story . . . and you get the whole theory of glaciation thrown in." Burroughs, who disliked arguing about anything, admitted that he

INTRODUCTION

often "used to chafe a good deal under [Muir's] biting Scotch wit and love of contradiction," but despite their differences, the two men became friends. Burroughs described Muir as "a poet and almost a Seer. . . . He could not sit down in a corner of the landscape as Thoreau did, he must have a continent for his playground."

It was with Muir, three years after the meeting at Riverby, that Burroughs traveled to Alaska as a member of the E. H. Harriman Expedition in 1899. Burroughs was then sixty-two, and the trip marked the first time he traveled west of the Mississippi. In the beginning of his 130-page account of the trip, "In Green Alaska" (excluded from this edition only because of its length), Burroughs establishes the agrarian values and eastern bias that influenced his reactions to the wilderness. "How staid and settled and old Nature looks in the Atlantic States, with her clear streams, her rounded hills, her forests, her lichen-covered rocks, her neutral tints, in contrast with large sections of the Rocky Mountain region." While he was crossing the continent, the farms he saw in the "bare, brown and forbidding" country of Utah, Idaho, and eastern Oregon affected him "like a nightmare." "The landscape suggests the dumping-ground of creation, where all the refuse has been gathered."

INTRODUCTION

Sailing from Victoria, however, Burroughs was able to notice "features that might have been gathered from the Highlands of the Hudson . . . with the addition of towering snow-capped peaks thrown in for background." At Muir Glacier he was moved and fascinated by the "largeness of the view, the elemental ruggedness, and the solitude as of interstellar space." But after his eyes became "sated with wild, austere grandeur," he was delighted to come into Ugak Bay and Kodiak Island where spruce forests, wildflowers, and the greenest grass he had ever seen impressed him as a "dream of rural beauty and repose."

In all his travels, Burroughs measured the wilderness in terms of what was familiar. Even when crossing the Atlantic he noted that the "only things that look familiar at sea are the clouds. These are messengers from home." And years later, when he was again traveling with Muir, the "home-like seclusion" of Yosemite Valley won his heart at once, initially because a robin, the "first I had seen since leaving home . . . brought the scene home to me, he supplied the link of association." His negative reactions to Jamaica are not so different from those he notes in his journal during a trip through the Deep South: "the scenery repels me. Nature here is too crude and watery and harsh; has not

INTRODUCTION

been subdued and tamed." In the mountains,
Burroughs describes the sense of loneliness and
insignificance that come to most people, even mountaineers, in the presence of an inhuman wilderness.
In his own backyard, near Slide Mountain and
Peakamoose in the Catskills, he could discover a
"confusion of the growth and decay of centuries"
that strikes him as "merciless and inhuman." Even
when gaining a glimpse of the opposite range of the
valley below, "one does not know his own farm or
settlement when framed in these mountain treetops;
all look alike unfamiliar."

Though always glad to get back to the familiar,
Burroughs still made frequent camping and fishing
trips, traveling into the Catskills and the Adirondacks, to Maine and to Canada, for a measured
taste of the wild. One of the reasons he gave for
building a cabin in the woods was that the area
around Riverby was growing too civilized for him
and he was attracted to the "wilderness and seclusion" of the muck swamp. Yet one of the first things
he did in the woods was to begin clearing the swamp
land and planting three acres of celery for market.
Burroughs was most attracted to areas having the
"wildness and freedom of nature, and marked by
those half-cultivated, half-wild features which birds
and boys love." Unlike John Muir, Burroughs was

less interested in a sublime panorama of wild nature than in a more intimate and personal one. One can't become intimate with a river, he said about his home on the Hudson, but in a secluded nook of the woods, as he shows in "Wild Life about My Cabin," one can come to know nature intimately and personally. "What one observes truly about bird or beast upon his farm of ten acres he will not have to unlearn, travel as wide and far as he will," he writes in "Animal Communication." "Your natural history knowledge of the East will avail you in the West."

A scientific impulse behind Burroughs's observations becomes evident in the title essay of this volume, "A Sharp Lookout." In some essays, there are images of Burroughs loitering, as he says, in cathedral aisles of hemlock woods, or wandering through fields at twilight, or sitting on a rock "with hands full of pink azalea, to listen." In other essays we see Burroughs more actively investigating nature, counting the number of times the hermit thrush sings per minute, or digging up a weasel's tunnels to discover its den, or timing the trips of a chipmunk transferring grain from one storage den to another. When Burroughs disclaims a scientific interest in nature, he is usually rejecting the science of the taxonomist and anatomist, the science of the

INTRODUCTION

laboratory and natural history museum when it offers only a "dead, dissected nature, a cabinet of curiosities carefully labeled and classified." He was always interested in life histories and habits in the wild; "what I want is light upon the whole of Nature—her methods, her laws, her results, her non-human ways . . . details are indispensable to the specialist, but a knowledge of relations and of wholes satisfies me more."

No doubt Burroughs's interests coincide with those of many early naturalists and scientists who have come to be identified with the history of ecology. Frequently misassociated solely with problems of pollution and the environment, ecology, properly speaking, studies the interactions of organisms with their environments. Although such study has a long history and the term *Oekologie* was coined in 1866 by the German biologist Ernst Haeckel, the first stirrings of ecology as a recognizable science did not occur until the end of the nineteenth century. In the United States, a zoologist, Stephen Forbes, produced an early classic of ecological theory in "The Lake as a Microcosm," published in 1887, but the self-conscious development of ecology as a formal science is more generally attributed to the work of plant ecologists. In the 1890s, ecologists were primarily seen as botanists who studied plant physiology outdoors. In 1893, Louis Pammel published *Flower Ecology*, the first American book with ecology in its title, and a group of botanists at the Madison Botanical Congress of 1893 adopted the word to define their new studies of plant adaptation and distribution.

Influenced by plant ecologists, particularly Frederic Clements and Henry Chandler Cowles, pioneer animal ecologists achieved some prominence in the second decade of the twentieth century. Charles Adams published his *Guide to the Study of Animal Ecology* in 1913, and an animal ecologist, Victor Shelford, became the first president of the Ecological Society of America in 1914. Often called the "new natural history" or "scientific natural history," ecology in these early years was not so much a coherent science as a set of general attitudes that cut across the disciplinary lines of several sciences. There was little agreement about specific concepts or methods. Two early influential concepts were developed in Victor Shelford's definition of ecology as the science of communities, and Frederic Clements's paradigm of a community as an organism, or more precisely, a superorganism.

Burroughs met some young ecologists on the Harriman Alaska Expedition in 1899. Two of them, Frederick Coville and Thomas Kearney, worked

INTRODUCTION

as botanists for the United States Department of
Agriculture. Under the influence of their mentor
and boss, C. Hart Merriam, who led the scientific
contingent of the Harriman Expedition, Frederick
Coville became best known for his ecological studies
of plant adaptation in Death Valley, and Thomas
Kearney for his "Study in the Ecology of North
Carolina Strand Vegetation." Both men extended
their ecological investigations beyond those of their
mentor Merriam, who maintained that the single
factor determining the geographical distribution of
species was temperature; Coville and Kearney included other factors, such as climate and soil, in
their studies of plant distribution and adaptation.
A third botanist on the expedition was the director
of the Missouri Botanical Garden, William Trelease, whose main ecological interest was in the
relationship between pollination and seasonal change.

After the Harriman Expedition, William Trelease
sent Burroughs some of his published work, and
Frederick Coville corresponded from Washington,
D.C., where he was curator of the National Her-
barium. In 1921, Coville sent Burroughs some
information about making syrup from sweet potatoes and noted how much he was enjoying the new
biography *John Burroughs, Boy and Man,* by Clara
Barrus. Burroughs, however, does not mention in

INTRODUCTION

his journals or essays the work of these young ecologists who began to publish during the last decades of his life. In a letter from Alaska he spoke of the "fearfully and wonderfully learned" scientists on the trip. "The Botanists and Zoologists talk in Latin most of the time, and Geologists have a jargon of their own. . . . Oh, these specialists, who cannot see the flower for its petals and stamens, or the mountain for its stratification!"

Although Burroughs shared a general interest with field botanists in studying relationships in the natural world, it is likely he would have found the work of some later animal ecologists more congenial to his own. At the Bard College symposium honoring Burroughs in 1978, the naturalist James Stapleton pointed out numerous instances where Burroughs seems to have anticipated findings later developed by animal ecologists. This is not to say that Burroughs was unique in his observations, but simply to say he was interested in making the kinds of observations that later animal ecologists considered worthy of investigation. For instance, Burroughs's observation of the relationship between the breeding habits of the ruby-crowned kinglet and the male's competitive display of its distinguishing patch of color relate to the more elaborate studies of birds' mating habits begun by Lorenz and Tinber-

INTRODUCTION

gen twenty years after Burroughs's death. His differentiation of birds according to feeding habits rather than morphological characteristics used by earlier scientists bears resemblance to the present practice of studying birds in terms of their "guilds" and "niches," how and where they live in the natural world. For example, Burroughs distinguished groups of warblers according to whether they take insects on the wing or pick them from the surface of bark and lichens, or probe deeply into the bark of trees as does the brown creeper. The difference he observed between the varying migration dates of early spring birds and the more precise dates of late migrating birds led to his speculations about the correlation between weather and biological activity, which is one of the concerns of the science that has come to be called phenology.

Burroughs's observations have been supported by studies showing that the return of birds in March is triggered by changes in the weather, while late April and May birds seem less affected by the weather and apparently return according to more predictable seasonal changes such as the lengthening daylight hours of spring. His careful observation of other animal and bird habits, such as modes of flocking or banding together, the maintenance of pecking orders, methods of signaling to each other, and variations of the homing instinct are important concerns in ethology, the scientific study of animal behavior. But Burroughs's studies were scientific only to a point; he did not develop theories or submit his intuitions to rigorous modes of testing and verification. Still, his habits of observation help us understand how he could say that "only a person with the scientific habit of mind can be trusted to report things as they are."

Burroughs's most vehement defense of scientific objectivity and accuracy occurs in a number of essays written early in the twentieth century, at "a time," as he said, "when birds and plants and trees are fast becoming a fad with half the population, and when the 'yellow' reporter is abroad in the fields and woods. Never before in my time have so many exaggerations and misconceptions of the wild life about us been current in the popular mind." In the pages of popular magazines such as *McClure's, Century, Harper's,* and *Country Life in America,* there appeared a new form of the nature essay in which humanized actions and thoughts of animals were presented as accurate natural history. Burroughs did not object to obvious fictional renditions of nature, such as those by Kipling, and he saw that a sentimental view of nature could have a good side by increasing our "feeling of kinship

INTRODUCTION

with all lower orders," but he objected to blatantly false interpretations of animal life when presented as scientific fact. Let the nature writer, he said, "make the most of what he sees, embellish it, amplify it, twirl it on the point of his pen like a juggler, but let him beware of adding to it; let him be sure he sees accurately."

Against charges of anthropomorphism in his own dramatization of nature, Burroughs would reply, "I am aware that it is my anthropomorphism that compels me to speak of Nature in this way; we have to describe that which is not man in terms of man, because we have no other." He goes on to say that bird songs are not music, properly speaking, but by writing them on the musical scale, "we only hint what we cannot express." Science, too, is a kind of anthropomorphism, Burroughs maintains. Terms such as gravity, osmosis, and electricity "stand for special activities in nature, and are as much inventions of our own minds as any of the rest of our ideas." But Burroughs does draw the line at the extent to which nature can be humanized, and he objected to the personification of wild creatures as "miniature human beings."

In 1903 Burroughs published "Real and Sham Natural History," an article primarily attacking the Reverend William J. Long's *School of the Woods*

INTRODUCTION

and Ernest Thompson Seton's *Wild Animals I Have Known,* which Burroughs retitled "Wild Animals I Alone Have Known" because, according to Burroughs, the book's wolves, foxes, rabbits, mustangs, and crows are animals that, "it is safe to say, no other person in the world has ever known." Burroughs scoffed at the portrayal of birds and animals consciously teaching their young as if in school, or acting in accord with the promptings of some moral code, or deliberately calculating defenses to prevent captivity. In a later essay, Burroughs distinguished ways of presenting human characteristics in animals:

They all have human traits and ways; let those be brought out—their mirth, their joy, their curiosity, their cunning, their thrift, their relations, their wars, their loves—and all the springs of their actions laid bare as far as possible; but I do not expect my natural history to back up the Ten Commandments, or to be an illustration of the value of training-schools and kindergartens, or to afford a commentary upon the vanity of human wishes. Humanize your facts to the extent of making them interesting, if you have the art to do it, but leave the dog a dog, and the straddle-bug a straddle-bug.

Despite his protests against "anthropomorphic absurdities," many readers were still attracted to Burroughs's own depiction of natural events as human dramas, and he did at times exceed his own restrictions against the personification of nature. At

INTRODUCTION

their best, however, his essays observe the distinction he makes between presenting the facts of nature with imagination and the imagining of facts themselves. In some of his earlier essays, Burroughs follows Emerson in saying that man can "have but one interest in nature, namely to see himself reflected or interpreted there." Later, he explains that facts about nature are sufficiently humanized the moment they become interesting, and by our human interest in wild creatures, he says simply, "I mean our interest in them as living, struggling beings."

To attribute human motives to the animals is to caricature them; but to put us in such relation with them that we feel their kinship, that we see their lives embosomed in the same iron necessity as our own, that we see in their minds a humbler manifestation of the same psychic power and intelligence that culminates and is conscious of itself in man,—that, I take it, is the true humanization.

The controversy Burroughs initiated eventually gained national attention. Theodore Roosevelt sent him a letter congratulating him on the article of 1903. During that same year Burroughs joined Roosevelt on a trip to Yellowstone, discovering during train stops that in some places he seemed as well known to the crowds as the President. In 1907 the natural history controversy reached the daily papers after Roosevelt himself joined the attack in "Men Who Misinterpret Nature" and "Nature Fakers," published in *Everybody's* magazine. In his own later essays, Burroughs continued to draw distinctions between writers of natural history and "unnatural natural history," but he softened his stance toward Ernest Thompson Seton, who, in turn, became more purposeful when presenting the facts of natural history. The two men became friends, and Burroughs found Seton to be a sound naturalist whose later work he admired, "although," as Burroughs wrote in a friendly letter, "if I find anything to criticize in it, you will hear from me." In 1926, Seton's four-volume work, *Lives of the Game Animals*, was awarded the John Burroughs Medal for distinguished nature writing.

Careful to avoid the pitfalls of the nature fakers, Burroughs began to investigate the similarities and differences between animals and human beings in such essays as "What Do Animals Know?," "Human Traits in the Animals," and "Animal and Plant Intelligence." Following T. H. Huxley, Burroughs states that what distinguishes humans from animals is language: "animals do not think in any proper sense, because they have no terms in which to think—no language." Yet animals do express pleasure or pain, fear or suspicion, and they do signal to each other through cries and calls. An astute student of bird songs all his life, Burroughs noted the relationship of birds' songs to their nesting habits, a connection supported by controlled experiments in recent years.

The most interesting of Burroughs's observations about animal communication, as James Stapleton has noted, still offer rich sources of investigation for ethologists. Burroughs's observations begin with what he calls the "life and death race," such as when a hawk will pursue a sparrow through a zigzag course and seemingly anticipate the sparrow's darting to the left or right without losing a stroke of the wing. The same is true of birds in their "love chasings" when the erratic flight of two birds is perfectly timed as though the pursuing bird knows the mind of the pursued. Or when a flock of snow buntings or blackbirds is on the alert, the flock, as if sprung by electricity, will take flight as though there were only one bird instead of a hundred. In "Animal Communication," Burroughs suggests that a "community of mind" exists among birds that is not present among humans, except, in occasional instances, when the pressure of anger, fear, or some other strong excitement will cause a group without leadership to behave like a single organism. No matter what explanations might be offered for this experience among men and women, there still remains for science to explain the more mysterious and fascinating movements of flocks of shore birds performing their startling evolutions in the air, "turning and flashing in the sun with a unity and precision that would be hard to imitate."

Darwin was the model of the great scientist for Burroughs. After reading *The Descent of Man,* Burroughs noted in his journal that the book "convinces like Nature herself. I have no more doubt of its main conclusions than I have of my own existence." But the best thing about Darwinism, as he often said, was Darwin himself: a "model of patient, tireless, sincere inquiry, [possessing] such candor, such love of truth, such keen insight into the methods of Nature, such singleness of purpose, and such nobility of mind." The most minute facts engaged Darwin. He counted nine thousand seeds, one by one, from artificially fertilized pods. He spent eight years dissecting barnacles, "but he was Darwin, and did not stop at barnacles . . . the upshot of his work was a tremendous gain to our understanding of the universe." In contrast, when Burroughs one day saw "a lot of college girls dissecting cats and making diagrams of the circulation and muscle-attachments, . . . I thought it pretty poor business unless the girls were taking a course in comparative anatomy with a view to some occupation in life. . . . I would rather see the girls in the fields and woods studying and enjoying living nature, training their eyes to see correctly and their hearts to respond intelligently." Seeing correctly, as Burroughs says in the same essay, is evident in Darwin whose "sympathies were so large and comprehensive" that he was determined "to see things, facts, in their relations."

Burroughs took great pleasure in the world of new truths that science offers us. "I graze eagerly in every one of its fields—astronomy, geology, botany, zoology, physics, chemistry, natural history"; but he added, "I do little more than graze in these fields." He was always intensely aware of the limitations of science. In numerous essays he lists facts and observations that neither laboratory scientists nor Darwin had explained or even investigated. Along with these smaller mysteries and secrets of the natural world, science has left intact the larger mysteries that became the concern of Burroughs's philosophical reflections in his late books. He notes that natural selection is not a creative process, but is rather a weeding-out process and plays only a negative part in the origin of species. "Whence the impulse that sent man forward?" he asks. "Why was one animal form endowed with the capacity for endless growth and development and all others denied it. Ah! that is the question of questions."

Burroughs frequently notes that natural selection can account for the *survival* of the fittest, but not for the *arrival* of the fittest: "How the organic came to bud and grow from the inorganic, who knows?" This problem constituted the "final mystery" for Burroughs, and he could not accept a mechanistic explanation or one that relied on a notion of "chance" any more than he could accept the religious accounts of his childhood. "An explanation of life phenomena that savors of the laboratory and chemism repels me, and an explanation that savors of the theological point of view is equally distasteful to me." Life is not just matter, and yet it cannot be separated from matter or the processes of the material world.

For a time, Burroughs found the gap in his thought filled by Bergson's *Creative Evolution* and the notion of *élan vital*, enabling him to view evolution as directed by a creative energy. As Burroughs had often said in earlier essays, it is not light passively received that creates the eye; rather, it is light meeting an "indwelling need in the organism, which amounts to an active creative principle, that begets the eye." The world becomes a blending of spirit and matter, even though the intellect cannot comprehend how they do blend. Toward the end of his life Burroughs found himself again trying to read Bergson's *Creative Evolution,* but this time "with poor success" discovering the work to be a "mixture of two things that won't mix—metaphysics and natural science. It is full of word-splitting and conjuring with terms" and the "logic is not strong." In the end, the "final mystery cannot be cleared up. We can only drive it to cover."

Burroughs was aware of the contradictions and inconsistencies in his work and life; as the student of Emerson and the friend of Whitman, he could live comfortably with them. He rejected the details of much scientific investigation, but carefully timed the songs of birds and weighed the nearly weightless bodies of hornets and cicadas. He serenely accepted the cruelties of nature yet grieved deeply over the deaths of his dogs. He attacked writers who humanized animals while continuing to personify birds himself. In one essay he states that nature is a hieroglyph to be translated, and in another he says it is not a cryptograph to be deciphered. Nature must be seen transcendentally, he says at one point, while stressing at another that it is enough to see nature with the naked eye. He praises some hunters as knowing nature more deeply than scientists while condemning others as murderers. At times he claims that science redeems the world; at other times he says it sterilizes the world.

Some of Burroughs's contradictory positions, as we have seen, can best be understood in relation to his formative experience as a farm boy. The conflict between civilization and nature, which is a major theme for most nature writers, did not preoccupy Burroughs to a great extent. He saw routines such as clearing and plowing fields, burning stumps, and tapping maples as activities that characterize a basic relationship of man to nature.

Hunting, too, was a natural part of life to rural man, and early in his career Burroughs advised a young woman interested in ornithology, "As for shooting the birds, I think a real lover of nature will indulge in no sentimentalism on the subject. Shoot them, of course, and no toying about it." Burroughs himself shot birds for ornithological purposes, and he once hunted deer by jacklight in the Adirondacks. He defended Roosevelt's big-game hunting as enabling the President to know the animals better than Burroughs himself did. His views did change, however, and he later advised ornithologists to leave their guns at home. In a letter to his son Julian in 1900, he wrote, "I saw the trail of the quails, poor little things, six of them now . . . real babes in the woods. And yet you can kill them!" He raged against plume hunters and indiscriminate egg collectors, and in "The Ways of Sportsmen," he said that a "man in the woods, with a gun in his hand, is no longer a man,—he is a brute." Yet in the journals, we find Burroughs, eighty years old, still blasting away at woodchucks.

Burroughs never became an active conservationist in the strict sense of the word. With the exception of his attacks on the nature fakers, he avoided controversies of all kinds. He was nevertheless a true conservationist in the sense described by Edwin Way Teale as one whose deep love and appreciation of wild things for their own sake form the only enduring component of any conservation movement: "It is only those who are deeply and fundamentally interested in nature itself who, in the long haul, the all-important continuity of effort, carry on." Burroughs accepted the fact that man disturbs the balance of nature wherever he goes, adding to the supply of food for some species while cutting off the supply for others. But the imbalance man creates while making his place in the world was becoming more extreme and alarming. "Where there is no vision," Burroughs writes, "science will not save us. In such a case our civilization is like an engine running without a headlight."

INTRODUCTION

Unlike earlier man on the farm, man in the twentieth century has placed himself more directly in opposition to elemental nature. "We live in an age of iron and have all we can do to keep the iron from entering our souls." Burroughs wonders whether the time is coming when man's scientific knowledge and the "vast system of artificial things with which it has enabled him to surround himself [will] cut short his history upon the planet. Will Nature in the end be avenged for the secrets he has forced from her?" Thinking of our civilization's terrible wasting of forests, coal, oil, minerals, soil, and wildlife, Burroughs "cannot but reflect what a sucked orange the earth will be in the course of a few more centuries." In his old age, while visiting Pittsburgh, "the devil's laboratory," he imagines how much more attractive life will be when instead of seeking the coal and oil in the earth, industries come "above the surface for the white coal, for the smokeless oil, for the winds and the sunshine. . . . Our very minds ought to be cleaner."

In "The Grist of the Gods," Burroughs presents a visionary portrait of the earth and the "thin pellicle of soil with which the granite framework of the globe is clothed." And, as he says in another late essay, until we see our true relation to the forces amid which we live and move—our concrete

INTRODUCTION

bodily relations—we are like children playing with fire; "man's fate is bound up with the fate of the planet." A new emphasis comes into these late essays as Burroughs looks more intently at the apparently inert matter supporting life. Noting that biology is rooted in geology and that life flowers out of rock, he shifts his view from creatures to "friendly rocks" and "warm and animated soil." At one point Burroughs observes how science "takes down the protecting roof of the heavens above us and shows us an unspeakable void." We then feel the "cosmic chill," but at the same time it is science, especially astronomy and geology, "that has builded us a new house—builded it over our heads while we were yet living in the old." Burroughs likewise goes about building his own house—as indicated in the titles of late works—*Under the Apple Trees* and "Under Genial Skies." The facts of science, he says, "make me feel at home in the world," the "ground underneath becomes a history, the stars overhead a revelation." His home has moved from a secluded nook in the woods and becomes the world itself, but the wellspring of his feeling for such a home is never lost.

In the last essay he saw in print, "Under Genial Skies," Burroughs considers the ways of seals off the California coast. Just as he once considered the mystery of the homing instinct of birds, he now speculates about the ability of seals to congregate in the same trysting place day after day in the same seemingly unmarked spot of water in the Pacific. "What is the secret of it?" Burroughs, now in his eighties, still asks. In the early morning, while lying in bed, he can hear the barking of the seals off the coast, and he imagines these "sea-dogs" chasing their prey and barking at short intervals just "as did the foxhounds I heard on the Catskills in my youth." In 1921, five days before his eighty-fourth birthday, Burroughs died aboard a train while crossing Ohio on his return from California. His last words were the question that concerned him much of his life, "How far are we from home?"

THE RETURN OF THE BIRDS

John Burroughs

EDITOR'S NOTE

"The Return of the Birds" appeared in *The Atlantic Monthly* in May 1865 under the title "With the Birds." It was Burroughs's first published nature essay, and it attracted the interest of Emerson. In revising the essay, Burroughs worked to remove what he called his "affectation, or want of simplicity," and to depend on "closer observation." As he later told Emerson, he removed a criticism of Thoreau's faulty observation in northern Maine because he himself had not been to the Maine woods. He significantly revised the opening pages to emphasize the organizational structure of the essay upon a seasonal motif, ranging from the cyclical return of the first migrating birds in early March to the departure of the last hawk in late autumn. The essay underscores Burroughs's interest in delineating the individual characteristics of birds while simultaneously dramatizing the general processes of nature and the relationships of species to seasonal change and natural habitats. It is interesting to note his allusion to Darwin in this early essay, although his own speculations on evolution-

EDITOR'S NOTE

ary theory at this point are more Lamarckian than Darwinian. In the most dramatic scene of the essay, however, Burroughs both embraces and evades a traditional view of Darwinian nature when he confronts the "terrible beauty" and "almost winged locomotion" of a black snake eating a baby catbird. Begun in 1863, the essay is reprinted here in its final form, published as the initial chapter of Burroughs's first nature book, *Wake-Robin*, in 1871.

F.B.

I

THE RETURN OF THE BIRDS

SPRING in our northern climate may fairly be
said to extend from the middle of March to
the middle of June. At least, the vernal tide continues to rise until the latter date, and it is not till
after the summer solstice that the shoots and
twigs begin to harden and turn to wood, or the
grass to lose any of its freshness and succulency.

It is this period that marks the return of the
birds, — one or two of the more hardy or half-
domesticated species, like the song sparrow and
the bluebird, usually arriving in March, while the
rarer and more brilliant wood-birds bring up the
procession in June. But each stage of the advancing season gives prominence to certain species, as
to certain flowers. The dandelion tells me when to
look for the swallow, the dogtooth violet when to
expect the wood-thrush, and when I have found the
wake-robin in bloom I know the season is fairly
inaugurated. With me this flower is associated, not
merely with the awakening of Robin, for he has

been awake some weeks, but with the universal awakening and rehabilitation of nature.

Yet the coming and going of the birds is more or less a mystery and a surprise. We go out in the morning, and no thrush or vireo is to be heard ; we go out again, and every tree and grove is musical; yet again, and all is silent. Who saw them come ? Who saw them depart ?

This pert little winter wren, for instance, darting in and out the fence, diving under the rubbish here and coming up yards away, — how does he manage with those little circular wings to compass degrees and zones, and arrive always in the nick of time ? Last August I saw him in the remotest wilds of the Adirondacks, impatient and inquisitive as usual ; a few weeks later, on the Potomac, I was greeted by the same hardy little busybody. Does he travel by easy stages from bush to bush and from wood to wood ? or has that compact little body force and courage to brave the night and the upper air, and so achieve leagues at one pull ?

And yonder bluebird with the earth tinge on his breast and the sky tinge on his back, — did he come down out of heaven on that bright March morning when he told us so softly and plaintively that, if we pleased, spring had come ? Indeed, there is nothing in the return of the birds more curious and suggestive than in the first appearance, or rumors of the appearance, of this little blue-coat. The bird at first seems a mere wandering voice in the air : one hears its call or carol on some bright March morning, but is uncertain of its source or direction ; it falls like a drop of rain when no cloud is visible ; one looks and listens, but to no purpose. The weather changes, perhaps a cold snap with snow comes on, and it may be a week before I hear the note again, and this time or the next perchance see the bird sitting on a stake in the fence lifting his wing as he calls cheerily to his mate. Its notes now become daily more frequent; the birds multiply, and, flitting from point to point, call and warble more confidently and gleefully. Their boldness increases till one sees them hovering with a saucy, inquiring air about barns and out-buildings, peeping into dove-cotes and stable windows, inspecting knotholes and pump-trees, intent only on a place to nest. They wage war against robins and wrens, pick quarrels with swallows, and seem to deliberate for days over the policy of taking forcible possession of one of the mud-houses of the latter. But as the season advances they drift more into the background. Schemes of conquest which they at first seemed bent upon are abandoned, and they settle down very quietly in their old quarters in remote stumpy fields.

Not long after the bluebird comes the robin, sometimes in March, but in most of the Northern

States April is the month of the robin. In large numbers they scour the fields and groves. You hear their piping in the meadow, in the pasture, on the hillside. Walk in the woods, and the dry leaves rustle with the whir of their wings, the air is vocal with their cheery call. In excess of joy and vivacity, they run, leap, scream, chase each other through the air, diving and sweeping among the trees with perilous rapidity.

In that free, fascinating, half-work and half-play pursuit, — sugar-making, — a pursuit which still lingers in many parts of New York, as in New England, — the robin is one's constant companion. When the day is sunny and the ground bare, you meet him at all points and hear him at all hours. At sunset, on the tops of the tall maples, with look heavenward, and in a spirit of utter abandonment, he carols his simple strain. And sitting thus amid the stark, silent trees, above the wet, cold earth, with the chill of winter still in the air, there is no fitter or sweeter songster in the whole round year. It is in keeping with the scene and the occasion. How round and genuine the notes are, and how eagerly our ears drink them in! The first utterance, and the spell of winter is thoroughly broken, and the remembrance of it afar off.

Robin is one of the most native and democratic of our birds; he is one of the family, and seems much nearer to us than those rare, exotic visitants.

as the orchard starling or rose-breasted grosbeak, with their distant, high-bred ways. Hardy, noisy, frolicsome, neighborly, and domestic in his habits, strong of wing and bold in spirit, he is the pioneer of the thrush family, and well worthy of the finer artists whose coming he heralds and in a measure prepares us for.

I could wish Robin less native and plebeian in one respect, — the building of his nest. Its coarse material and rough masonry are creditable neither to his skill as a workman nor to his taste as an artist. I am the more forcibly reminded of his deficiency in this respect from observing yonder hummingbird's nest, which is a marvel of fitness and adaptation, a proper setting for this winged gem, — the body of it composed of a white, felt-like substance, probably the down of some plant or the wool of some worm, and toned down in keeping with the branch on which it sits by minute tree-lichens, woven together by threads as fine and frail as gossamer. From Robin's good looks and musical turn, we might reasonably predict a domicile of equal fitness and elegance. At least I demand of him as clean and handsome a nest as the king-bird's, whose harsh jingle, compared with Robin's evening melody, is as the clatter of pots and kettles beside the tone of a flute. I love his note and ways better even than those of the orchard starling or the Baltimore oriole ; yet his nest, compared with theirs, is a half-subterranean hut contrasted with a Roman villa. There is something courtly and poetical in a pensile nest. Next to a castle in the air is a dwelling suspended to the slender branch of a tall tree, swayed and rocked forever by the wind. Why need wings be afraid of falling? Why build only where boys can climb? After all, we must set it down to the account of Robin's democratic turn : he is no aristocrat, but one of the people ; and therefore we should expect stability in his workmanship, rather than elegance.

Another April bird, which makes her appearance sometimes earlier and sometimes later than Robin, and whose memory I fondly cherish, is the phœbe-bird, the pioneer of the flycatchers. In the inland farming districts, I used to notice her, on some bright morning about Easter Day, proclaiming her arrival, with much variety of motion and attitude, from the peak of the barn or hay-shed. As yet, you may have heard only the plaintive, homesick note of the bluebird, or the faint trill of the song sparrow; and Phœbe's clear, vivacious assurance of her veritable bodily presence among us again is welcomed by all ears. At agreeable intervals in her lay she describes a circle or an ellipse in the air, ostensibly prospecting for insects, but really, I suspect, as an artistic flourish, thrown in to make up in some way for the deficiency of her musical performance. If plainness of dress indicates powers of song, as it usually does, then Phœbe ought to be unrivaled in musical ability, for surely that ashen-gray suit is the superlative of plainness ; and that form, likewise, would hardly pass for a "perfect figure" of a bird. The seasonableness of her coming, however, and her civil, neighborly ways, shall make up for all deficiencies in song and plumage. After a few weeks Phœbe is seldom seen, except as she darts from her moss-covered nest beneath some bridge or shelving cliff.

Another April comer, who arrives shortly after Robin-redbreast, with whom he associates both at this season and in the autumn, is the gold-winged woodpecker, *alias* "high-hole," *alias* "flicker," *alias* "yarup." He is an old favorite of my boyhood, and his note to me means very much. He announces his arrival by a long, loud call, repeated from the dry branch of some tree, or a stake in the fence, — a thoroughly melodious April sound. I think how Solomon finished that beautiful description of spring, "And the voice of the turtle is heard in the land," and see that a description of spring in this farming country, to be equally characteristic, should culminate in like manner, — "And the call of the high-hole comes up from the wood."

It is a loud, strong, sonorous call, and does not seem to imply an answer, but rather to subserve some purpose of love or music. It is " Yarup's " proclamation of peace and good-will to all. On looking at the matter closely, I perceive that most birds, not denominated songsters, have, in the spring, some note or sound or call that hints of a song, and answers imperfectly the end of beauty and art. As a "livelier iris changes on the burnished dove," and the fancy of the young man turns lightly to thoughts of his pretty cousin, so the same renewing spirit touches the "silent singers," and they are no longer dumb; faintly they lisp the first syllables of the marvelous tale. Witness the clear, sweet whistle of the gray-crested titmouse, — the soft, nasal piping of the nuthatch, — the amorous, vivacious warble of the bluebird, — the long, rich note of the meadowlark, — the whistle of the quail, — the drumming of the partridge, — the animation and loquacity of the swallows, and the like. Even the hen has a homely, contented carol; and I credit the owls with a desire to fill the night with music. All birds are incipient or would-be songsters in the spring. I find corroborative evidence of this even in the crowing of the cock. The flowering of the maple is not so obvious as that of the magnolia; nevertheless, there is actual inflorescence.

Few writers award any song to that familiar little sparrow, the *Socialis;* yet who that has observed him sitting by the wayside, and repeating, with devout attitude, that fine sliding chant, does

not recognize the neglect? Who has heard the
snowbird sing? Yet he has a lisping warble very
savory to the ear. I have heard him indulge in it
even in February.

Even the cow bunting feels the musical tendency,
and aspires to its expression, with the rest. Perched
upon the topmost branch beside his mate or mates,
— for he is quite a polygamist, and usually has
two or three demure little ladies in faded black beside him, — generally in the early part of the day,
he seems literally to vomit up his notes. Apparently with much labor and effort, they gurgle and
blubber up out of him, falling on the ear with
a peculiar subtile ring, as of turning water from
a glass bottle, and not without a certain pleasing
cadence.

Neither is the common woodpecker entirely in-
sensible to the wooing of the spring, and, like the
partridge, testifies his appreciation of melody after
quite a primitive fashion. Passing through the
woods on some clear, still morning in March, while
the metallic ring and tension of winter are still in
the earth and air, the silence is suddenly broken
by long, resonant hammering upon a dry limb or
stub. It is Downy beating a reveille to spring. In
the utter stillness and amid the rigid forms we listen
with pleasure; and, as it comes to my ear oftener
at this season than at any other, I freely exonerate
the author of it from the imputation of any gas-

tronomic motives, and credit him with a genuine musical performance.

It is to be expected, therefore, that "yellow-hammer" will respond to the general tendency, and contribute his part to the spring chorus. His April call is his finest touch, his most musical expression.

I recall an ancient maple standing sentry to a large sugar-bush, that, year after year, afforded protection to a brood of yellow-hammers in its decayed heart. A week or two before the nesting seemed actually to have begun, three or four of these birds might be seen, on almost any bright morning, gamboling and courting amid its decayed branches. Sometimes you would hear only a gentle persuasive cooing, or a quiet confidential chattering, — then that long, loud call, taken up by first one, then another, as they sat about upon the naked limbs, — anon, a sort of wild, rollicking laughter, intermingled with various cries, yelps, and squeals, as if some incident had excited their mirth and ridicule. Whether this social hilarity and boisterousness is in celebration of the pairing or mating ceremony, or whether it is only a sort of annual "house-warming" common among high-holes on resuming their summer quarters, is a question upon which I reserve my judgment.

Unlike most of his kinsmen, the golden-wing prefers the fields and the borders of the forest to the deeper seclusion of the woods, and hence, contrary to the habit of his tribe, obtains most of his subsistence from the ground, probing it for ants and crickets. He is not quite satisfied with being a woodpecker. He courts the society of the robin and the finches, abandons the trees for the meadow, and feeds eagerly upon berries and grain. What may be the final upshot of this course of living is a question worthy the attention of Darwin. Will his taking to the ground and his pedestrian feats result in lengthening his legs, his feeding upon berries and grains subdue his tints and soften his voice, and his associating with Robin put a song into his heart?

Indeed, what would be more interesting than the history of our birds for the last two or three centuries? There can be no doubt that the presence of man has exerted a very marked and .friendly influence upon them, since they so multiply in his society. The birds of California, it is said, were mostly silent till after its settlement, and I doubt if the Indians heard the wood thrush as we hear him. Where did the bobolink disport himself before there were meadows in the North and rice-fields in the South? Was he the same lithe, merry-hearted beau then as now? And the sparrow, the lark, and the goldfinch, birds that seem so indigenous to the open fields and so averse to the woods, — we cannot conceive of their existence in a vast wilderness and without man.

But to return. The song sparrow, that universal favorite and firstling of the spring, comes before April, and its simple strain gladdens all hearts.

May is the month of the swallows and the orioles. There are many other distinguished arrivals, indeed nine tenths of the birds are here by the last week in May, yet the swallows and orioles are the most conspicuous. The bright plumage of the latter seems really like an arrival from the tropics. I see them dash through the blossoming trees, and all the forenoon hear their incessant warbling and wooing. The swallows dive and chatter about the barn, or squeak and build beneath the eaves; the partridge drums in the fresh sprouting woods; the long, tender note of the meadowlark comes up from the meadow; and at sunset, from every marsh and pond come the ten thousand voices of the hylas. May is the transition month, and exists to connect April and June, the root with the flower.

With June the cup is full, our hearts are satisfied, there is no more to be desired. The perfection of the season, among other things, has brought the perfection of the song and plumage of the birds. The master artists are all here ; and the expectations excited by the robin and the song sparrow are fully justified. The thrushes have all come ; and I sit down upon the first rock, with hands full of the pink azalea, to listen. With me, the cuckoo does not arrive till June ; and often the goldfinch, the kingbird, the scarlet tanager delay their coming till then. In the meadows the bobolink is in all his glory; in the high pastures the field sparrow sings his breezy vesper-hymn ; and the woods are unfolding to the music of the thrushes.

The cuckoo is one of the most solitary birds of our forests, and is strangely tame and quiet, appearing equally untouched by joy or grief, fear or anger. Something remote seems ever weighing upon his mind. His note or call is as of one lost or wandering, and to the farmer is prophetic of rain. Amid the general joy and the sweet assurance of things, I love to listen to the strange clairvoyant call. Heard a quarter of a mile away, from out the depths of the forest, there is something peculiarly weird and monkish about it. Wordsworth's lines upon the European species apply equally well to ours : —

"O blithe new-comer ! I have heard,
  I hear thee and rejoice :
O cuckoo ! shall I call thee bird ?
  Or but a wandering voice ?

"While I am lying on the grass,
  Thy loud note smites my ear !
From hill to hill it seems to pass,
  At once far off and near !

WAKE–ROBIN

"Thrice welcome, darling of the spring!
Even yet thou art to me
No bird, but an invisible thing,
A voice, a mystery."

The black-billed is the only species found in my locality, the yellow-billed abounds farther south. Their note or call is nearly the same. The former sometimes suggests the voice of a turkey. The call of the latter may be suggested thus: *k-k-k-k-k-kow, kow, kow-ow, kow-ow*.

The yellow-billed will take up his stand in a tree, and explore its branches till he has caught every worm. He sits on a twig, and with a peculiar swaying movement of his head examines the surrounding foliage. When he discovers his prey, he leaps upon it in a fluttering manner.

In June the black-billed makes a tour through the orchard and garden, regaling himself upon the canker-worms. At this time he is one of the tamest of birds, and will allow you to approach within a few yards of him. I have even come within a few feet of one without seeming to excite his fear or suspicion. He is quite unsophisticated, or else royally indifferent.

The plumage of the cuckoo is a rich glossy brown, and is unrivaled in beauty by any other neutral tint with which I am acquainted. It is also remarkable for its firmness and fineness.

Notwithstanding the disparity in size and color,

the black-billed species has certain peculiarities that remind one of the passenger pigeon. His eye, with its red circle, the shape of his head, and his motions on alighting and taking flight, quickly suggest the resemblance; though in grace and speed, when on the wing, he is far inferior. His tail seems disproportionately long, like that of the red thrush, and his flight among the trees is very still, contrasting strongly with the honest clatter of the robin or pigeon.

Have you heard the song of the field sparrow? If you have lived in a pastoral country with broad upland pastures, you could hardly have missed him. Wilson, I believe, calls him the grass finch, and was evidently unacquainted with his powers of song. The two white lateral quills in his tail, and his habit of running and skulking a few yards in advance of you as you walk through the fields, are sufficient to identify him. Not in meadows or orchards, but in high, breezy pasture-grounds, will you look for him. His song is most noticeable after sundown, when other birds are silent; for which reason he has been aptly called the vesper sparrow. The farmer following his team from the field at dusk catches his sweetest strain. His song is not so brisk and varied as that of the song sparrow, being softer and wilder, sweeter and more plaintive. Add the best parts of the lay of the latter to the sweet vibrating chant of the wood sparrow, and you have the evening hymn of the vesper-bird, — the poet of the plain, unadorned pastures. Go to those broad, smooth, uplying fields where the cattle and sheep are grazing, and sit down in the twilight on one of those warm, clean stones, and listen to this song. On every side, near and remote, from out the short grass which the herds are cropping, the strain rises. Two or three long, silver notes of peace and rest, ending in some subdued trills and quavers, constitute each separate song. Often you will catch only one or two of the bars, the breeze having blown the minor part away. Such unambitious, quiet, unconscious melody! It is one of the most characteristic sounds in nature. The grass, the stones, the stubble, the furrow, the quiet herds, and the warm twilight among the hills, are all subtly expressed in this song; this is what they are at last capable of.

The female builds a plain nest in the open field, without so much as a bush or thistle or tuft of grass to protect it or mark its site; you may step upon it, or the cattle may tread it into the ground. But the danger from this source, I presume, the bird considers less than that from another. Skunks and foxes have a very impertinent curiosity, as Finchie well knows; and a bank or hedge, or a rank growth of grass or thistles, that might promise protection and cover to mouse or bird, these cunning rogues would be apt to explore most thoroughly. The partridge is undoubtedly acquainted with the same process of reasoning ; for, like the vesper-bird, she, too, nests in open, unprotected places, avoiding all show of concealment, — coming from the tangled and almost impenetrable parts of the forest to the clean, open woods, where she can command all the approaches and fly with equal ease in any direction.

Another favorite sparrow, but little noticed, is the wood or bush sparrow, usually called by the ornithologists *Spizella pusilla*. Its size and form is that of the *socialis*, but is less distinctly marked, being of a duller redder tinge. He prefers remote bushy heathery fields, where his song is one of the sweetest to be heard. It is sometimes very noticeable, especially early in spring. I remember sitting one bright day in the still leafless April woods, when one of these birds struck up a few rods from me, repeating its lay at short intervals for nearly an hour. It was a perfect piece of wood-music, and was of course all the more noticeable for being projected upon such a broad unoccupied page of silence. Its song is like the words, *fe-o, fe-o, fe-o, few, few, few, fee fee fee*, uttered at first high and leisurely, but running very rapidly toward the close, which is low and soft.

Still keeping among the unrecognized, the white-eyed vireo, or flycatcher, deserves particular mention. The song of this bird is not particularly sweet and soft ; on the contrary, it is a little hard and shrill, like that of the indigo-bird or oriole ; but for brightness, volubility, execution, and power of imitation, he is unsurpassed by any of our northern birds. His ordinary note is forcible and emphatic, but, as stated, not especially musical; *Chick-a-re'r-chick*, he seems to say, hiding himself in the low, dense undergrowth, and eluding your most vigilant search, as if playing some part in a game. But in July or August, if you are on good terms with the sylvan deities, you may listen to a far more rare and artistic performance. Your first impression will be that that cluster of azalea, or that clump of swamp-huckleberry, conceals three or four different songsters, each vying with the others to lead the chorus. Such a medley of notes, snatched from half the songsters of the field and forest, and uttered with the utmost clearness and rapidity, I am sure you cannot hear short of the haunts of the genuine mockingbird. If not fully and accurately repeated, there are at least suggested the notes of the robin, wren, catbird, high-hole, goldfinch, and song sparrow. The *pip*, *pip*, of the last is produced so accurately that I verily believe it would deceive the bird herself; and the whole uttered in such rapid succession that it seems as if the movement that gives the concluding note of one strain must form the first note of the next. The effect is very rich, and, to my ear, entirely unique. The performer is very careful not to reveal himself in the mean time; yet there is a conscious air about the strain that impresses me with the idea that my presence is understood and my attention courted. A tone of pride and glee, and, occasionally, of bantering jocoseness, is discernible. I believe it is only rarely, and when he is sure of his audience, that he displays his parts in this manner. You are to look for him, not in tall trees or deep forests, but in low, dense shrubbery about wet places, where there are plenty of gnats and mosquitoes.

The winter wren is another marvelous songster, in speaking of whom it is difficult to avoid superlatives. He is not so conscious of his powers and so ambitious of effect as the white-eyed flycatcher, yet you will not be less astonished and delighted on hearing him. He possesses the fluency and copiousness for which the wrens are noted, and besides these qualities, and what is rarely found conjoined with them, a wild, sweet, rhythmical cadence that holds you entranced. I shall not soon forget that perfect June day, when, loitering in a low, ancient hemlock wood, in whose cathedral aisles the coolness and freshness seems perennial, the silence was suddenly broken by a strain so rapid and gushing, and touched with such a wild, sylvan plaintiveness, that I listened in amazement. And so shy and coy was the little minstrel, that I came twice to the woods before I was sure to whom I was listening. In summer he is one of those birds of the deep northern forests, that, like the speckled Canada warbler and the hermit thrush, only the privileged ones hear.

The distribution of plants in a given locality is not more marked and defined than that of the birds. Show a botanist a landscape, and he will tell you where to look for the lady's-slipper, the columbine, or the harebell. On the same principles the ornithologist will direct you where to look for the greenlets, the wood sparrow, or the chewink. In adjoining counties, in the same latitude, and equally inland, but possessing a different geological formation and different forest-timber, you will observe quite a different class of birds. In a land of the beech and sugar maple I do not find the same songsters that I know where thrive the oak, chestnut, and laurel. In going from a district of the Old Red Sandstone to where I walk upon the old Plutonic Rock, not fifty miles distant, I miss in the woods the veery, the hermit thrush, the chestnut-sided warbler, the blue-backed warbler, the green-backed warbler, the black and yellow warbler, and many others, and find in their stead the wood thrush, the chewink, the redstart, the yellow-throat, the yellow-breasted flycatcher, the white-eyed flycatcher, the quail, and the turtle dove.

In my neighborhood here in the Highlands the distribution is very marked. South of the village I invariably find one species of birds, north of it another. In only one locality, full of azalea and swamp-huckleberry, I am always sure of finding the hooded warbler. In a dense undergrowth of spice-bush, witch-hazel, and alder, I meet the worm-eating warbler. In a remote clearing, covered with heath and fern, with here and there a chestnut and an oak, I go to hear in July the wood sparrow, and returning by a stumpy, shallow pond, I am sure to find the water-thrush.

Only one locality within my range seems to possess attractions for all comers. Here one may study almost the entire ornithology of the State. It is a rocky piece of ground, long ago cleared, but now fast relapsing into the wildness and freedom of nature, and marked by those half-cultivated, half-wild features which birds and boys love. It is bounded on two sides by the village and highway, crossed at various points by carriage-roads, and threaded in all directions by paths and byways, along which soldiers, laborers, and truant school-boys are passing at all hours of the day. It is so far escaping from the axe and the bush-hook as to have opened communication with the forest and mountain beyond by straggling lines of cedar, laurel, and blackberry. The ground is mainly occupied with cedar and chestnut, with an undergrowth, in many places, of heath and bramble. The chief feature, however, is a dense growth in the centre, consisting of dogwood, water-beech, swamp-ash, alder, spice-bush, hazel, etc., with a network of smilax and frost-grape. A little zigzag stream, the draining of a swamp beyond, which passes through this tanglewood, accounts for many of its features and productions, if not for its entire existence. Birds that are not attracted by the heath, or the cedar and chestnut, are sure to find some excuse for visiting this miscellaneous growth in the centre. Most of the common birds literally throng this idle-wild; and I have met here many of the rarer species, such as the great-crested flycatcher, the solitary warbler, the blue-winged swamp warbler, the worm-eating warbler, the fox sparrow, etc. The absence of all birds of prey, and the great number of flies and insects, both the result of proximity to the village, are considerations which no hawk-fearing, peace-loving minstrel passes over lightly; hence the popularity of the resort.

But the crowning glory of all these robins, flycatchers, and warblers is the wood thrush. More abundant than all other birds, except the robin and catbird, he greets you from every rock and shrub. Shy and reserved when he first makes his appearance in May, before the end of June he is tame and familiar, and sings on the tree over your head, or on the rock a few paces in advance. A pair even built their nest and reared their brood within ten or twelve feet of the piazza of a large summer-house in the vicinity. But when the guests commenced to arrive and the piazza to be thronged with gay crowds, I noticed something like dread and foreboding in the manner of the mother bird; and from her still, quiet ways, and habit of sitting long and silently within a few feet of the precious charge, it seemed as if the dear creature had resolved, if possible, to avoid all observation.

If we take the quality of melody as the test, the wood thrush, hermit thrush, and the veery thrush stand at the head of our list of songsters.

The mockingbird undoubtedly possesses the greatest range of mere talent, the most varied executive ability, and never fails to surprise and delight one anew at each hearing ; but being mostly an imitator, he never approaches the serene beauty and sublimity of the hermit thrush. The word that best expresses my feelings, on hearing the mockingbird, is admiration, though the first emotion is one of surprise and incredulity. That so many and such various notes should proceed from one throat is a marvel, and we regard the performance with feelings akin to those we experience on witnessing the astounding feats of the athlete or gymnast,— and this, notwithstanding many of the notes imitated have all the freshness and sweetness of the originals. The emotions excited by the songs of these thrushes belong to a higher order, springing as they do from our deepest sense of the beauty and harmony of the world.

WAKE-ROBIN

The wood thrush is worthy of all, and more than all, the praises he has received; and considering the number of his appreciative listeners, it is not a little surprising that his relative and equal, the hermit thrush, should have received so little notice. Both the great ornithologists, Wilson and Audubon, are lavish in their praises of the former, but have little or nothing to say of the song of the latter. Audubon says it is sometimes agreeable, but evidently has never heard it. Nuttall, I am glad to find, is more discriminating, and does the bird fuller justice.

It is quite a rare bird, of very shy and secluded habits, being found in the Middle and Eastern States, during the period of song, only in the deepest and most remote forests, usually in damp and swampy localities. On this account the people in the Adirondack region call it the "Swamp Angel." Its being so much of a recluse accounts for the comparative ignorance that prevails in regard to it.

The cast of its song is very much like that of the wood thrush, and a good observer might easily confound the two. But hear them together and the difference is quite marked: the song of the hermit is in a higher key, and is more wild and ethereal. His instrument is a silver horn which he winds in the most solitary places. The song of the wood thrush is more golden and leisurely. Its tone comes near to that of some rare stringed instrument. One feels that perhaps the wood thrush has more compass and power, if he would only let himself out, but on the whole he comes a little short of the pure, serene, hymn-like strain of the hermit.

Yet those who have heard only the wood thrush may well place him first on the list. He is truly a royal minstrel, and, considering his liberal distribution throughout our Atlantic seaboard, perhaps contributes more than any other bird to our sylvan melody. One may object that he spends a little too much time in tuning his instrument, yet his careless and uncertain touches reveal its rare compass and power.

He is the only songster of my acquaintance, excepting the canary, that displays different degrees of proficiency in the exercise of his musical gifts. Not long since, while walking one Sunday in the edge of an orchard adjoining a wood, I heard one that so obviously and unmistakably surpassed all his rivals, that my companion, though slow to notice such things, remarked it wonderingly; and with one accord we paused to listen to so rare a performer. It was not different in quality so much as in quantity. Such a flood of it! Such copiousness! Such long, trilling, accelerating preludes! Such sudden, ecstatic overtures would have intoxicated the dullest ear. He was really without a compeer, — a master-artist. Twice afterward I was conscious of having heard the same bird.

The wood thrush is the handsomest species of this family. In grace and elegance of manner he has no equal. Such a gentle, high-bred air, and such inimitable ease and composure in his flight and movement! He is a poet in very word and deed. His carriage is music to the eye. His performance of the commonest act, as catching a beetle, or picking a worm from the mud, pleases like a stroke of wit or eloquence. Was he a prince in the olden time, and do the regal grace and mien still adhere to him in his transformation? What a finely proportioned form! How plain, yet rich, his color,— the bright russet of his back, the clear white of his breast, with the distinct heart-shaped spots! It may be objected to Robin that he is noisy and demonstrative; he hurries away or rises to a branch with an angry note, and flirts his wings in ill-bred suspicion. The mavis, or red thrush, sneaks and skulks like a culprit, hiding in the densest alders; the catbird is a coquette and a flirt, as well as a sort of female Paul Pry; and the chewink shows his inhospitality by espying your movements like a Japanese. The wood thrush has none of these underbred traits. He regards me unsuspiciously, or avoids me with a noble reserve,— or, if I am quiet and incurious, graciously hops toward me, as if to pay his respects, or to make my acquaintance. I have passed under his nest within a few feet of his mate and brood, when he sat near by on a branch eying

me sharply, but without opening his beak; but
the moment I raised my hand toward his defenseless
household his anger and indignation were beautiful
to behold.

What a noble pride he has! Late one October,
after his mates and companions had long since gone
south, I noticed one for several successive days in
the dense part of this next-door wood, flitting noise-
lessly about, very grave and silent, as if doing pen-
ance for some violation of the code of honor. By
many gentle, indirect approaches, I perceived that
part of his tail-feathers were undeveloped. The
sylvan prince could not think of returning to court
in this plight, and so, amid the falling leaves
and cold rains of autumn, was patiently biding his
time.

The soft, mellow flute of the veery fills a place
in the chorus of the woods that the song of the vesper sparrow fills in the chorus of the fields. It has
the nightingale's habit of singing in the twilight,
as indeed have all our thrushes. Walk out toward
the forest in the warm twilight of a June day, and
when fifty rods distant you will hear their soft,
reverberating notes rising from a dozen different
throats.

It is one of the simplest strains to be heard,— as
simple as the curve in form, delighting from the
pure element of harmony and beauty it contains,
and not from any novel or fantastic modulation of

it,— thus contrasting strongly with such rollicking, hilarious songsters as the bobolink, in whom we are chiefly pleased with the tintinnabulation, the verbal and labial excellence, and the evident conceit and delight of the performer.

I hardly know whether I am more pleased or annoyed with the catbird. Perhaps she is a little too common, and her part in the general chorus a little too conspicuous. If you are listening for the note of another bird, she is sure to be prompted to the most loud and protracted singing, drowning all other sounds; if you sit quietly down to observe a favorite or study a new-comer, her curiosity knows no bounds, and you are scanned and ridiculed from every point of observation. Yet I would not miss her ; I would only subordinate her a little, make her less conspicuous.

She is the parodist of the woods, and there is ever a mischievous, bantering, half-ironical undertone in her lay, as if she were conscious of mimicking and disconcerting some envied songster. Ambitious of song, practicing and rehearsing in private, she yet seems the least sincere and genuine of the sylvan minstrels, as if she had taken up music only to be in the fashion, or not to be outdone by the robins and thrushes. In other words, she seems to sing from some outward motive, and not from inward joyousness. She is a good versifier, but not a great poet. Vigorous, rapid, copious, not without fine touches, but destitute of any high, serene melody, her performance, like that of Thoreau's squirrel, always implies a spectator.

There is a certain air and polish about her strain, however, like that in the vivacious conversation of a well-bred lady of the world, that commands respect. Her maternal instinct, also, is very strong, and that simple structure of dead twigs and dry grass is the centre of much anxious solicitude. Not long since, while strolling through the woods, my attention was attracted to a small densely grown swamp, hedged in with eglantine, brambles, and the everlasting smilax, from which proceeded loud cries of distress and alarm, indicating that some terrible calamity was threatening my sombre-colored minstrel. On effecting an entrance, which, however, was not accomplished till I had doffed coat and hat, so as to diminish the surface exposed to the thorns and brambles, and, looking around me from a square yard of terra firma, I found myself the spectator of a loathsome yet fascinating scene. Three or four yards from me was the nest, beneath which, in long festoons, rested a huge black snake; a bird two thirds grown was slowly disappearing between his expanded jaws. As he seemed unconscious of my presence, I quietly observed the proceedings. By slow degrees he compassed the bird about with his elastic mouth ; his head flattened, his neck writhed and swelled, and two or three undulatory movements of his glistening body finished the work. Then he cautiously raised himself up, his tongue flaming from his mouth the while, curved over the nest, and, with wavy, subtle motions, explored the interior. I can conceive of nothing more overpoweringly terrible to an unsuspecting family of birds than the sudden appearance above their domicile of the head and neck of this arch-enemy. It is enough to petrify the blood in their veins. Not finding the object of his search, he came streaming down from the nest to a lower limb, and commenced extending his researches in other directions, sliding stealthily through the branches, bent on capturing one of the parent birds. That a legless, wingless creature should move with such ease and rapidity where only birds and squirrels are considered at home, lifting himself up, letting himself down, running out on the yielding boughs, and traversing with marvelous celerity the whole length and breadth of the thicket, was truly surprising. One thinks of the great myth of the Tempter and the " cause of all our woe," and wonders if the Arch One is not now playing off some of his pranks before him. Whether we call it snake or devil matters little. I could but admire his terrible beauty, however; his black, shining folds, his easy, gliding movement, head erect, eyes glistening, tongue playing like subtle flame, and the invisible means of his almost winged locomotion.

The parent birds, in the mean while, kept up the most agonizing cry,— at times fluttering furiously about their pursuer, and actually laying hold of his tail with their beaks and claws. On being thus attacked, the snake would suddenly double upon himself and follow his own body back, thus executing a strategic movement that at first seemed almost to paralyze his victim and place her within his grasp. Not quite, however. Before his jaws could close upon the coveted prize the bird would tear herself away, and, apparently faint and sobbing, retire to a higher branch. His reputed powers of fascination availed him little, though it is possible that a frailer and less combative bird might have been held by the fatal spell. Presently, as he came gliding down the slender body of a leaning alder, his attention was attracted by a slight movement of my arm; eyeing me an instant, with that crouching, utter, motionless gaze which I believe only snakes and devils can assume, he turned quickly,— a feat which necessitated something like crawling over his own body,— and glided off through the branches, evidently recognizing in me a representative of the ancient parties he once so cunningly ruined. A few moments after, as he lay carelessly disposed in the top of a rank alder, trying to look as much like a crooked branch as his supple, shining form would admit, the old vengeance overtook him. I exercised my prerogative, and a well-directed missile, in the shape of a stone, brought him looping and writhing to the ground. After I had completed his downfall and quiet had been partially restored, a half-fledged member of the bereaved household came out from his hiding-place, and, jumping upon a decayed branch, chirped vigorously, no doubt in celebration of the victory.

Till the middle of July there is a general equilibrium; the tide stands poised; the holiday spirit is unabated. But as the harvest ripens beneath the long, hot days, the melody gradually ceases. The young are out of the nest and must be cared for, and the moulting season is at hand. After the cricket has commenced to drone his monotonous refrain beneath your window, you will not, till another season, hear the wood thrush in all his matchless eloquence. The bobolink has become careworn and fretful, and blurts out snatches of his song between his scolding and upbraiding, as you approach the vicinity of his nest, oscillating between anxiety for his brood and solicitude for his musical reputation. Some of the sparrows still sing, and occasionally across the hot fields, from a tall tree in the edge of the forest, comes the rich note of the scarlet tanager. This tropical-colored bird loves the hottest weather, and I hear him even in dog-days.

The remainder of the summer is the carnival of the swallows and flycatchers. Flies and insects, to any amount, are to be had for the catching; and the opportunity is well improved. See that sombre, ashen-colored pewee on yonder branch. A true sportsman he, who never takes his game at rest, but always on the wing. You vagrant fly, you purblind moth, beware how you come within his range! Observe his attitude, the curious movement of his head, his "eye in a fine frenzy rolling, glancing from heaven to earth, from earth to heaven."

His sight is microscopic and his aim sure. Quick as thought he has seized his victim and is back to his perch. There is no strife, no pursuit,— one fell swoop and the matter is ended. That little sparrow, as you will observe, is less skilled. It is the *Socialis*, and he finds his subsistence properly in various seeds and the larvæ of insects, though he occasionally has higher aspirations, and seeks to emulate the pewee, commencing and ending his career as a flycatcher by an awkward chase after a beetle or "miller." He is hunting around in the grass now, I suspect, with the desire to indulge this favorite whim. There! — the opportunity is afforded him. Away goes a little cream-colored meadow-moth in the most tortuous course he is capable of, and away goes *Socialis* in pursuit. The contest is quite comical, though I dare say it is serious enough to the moth. The chase continues for a few yards, when there is a sudden rush-

ing to cover in the grass,— then a taking to wing
again, when the search has become too close, and
the moth has recovered his wind. *Socialis* chirps
angrily, and is determined not to be beaten. Keeping, with the slightest effort, upon the heels of the
fugitive, he is ever on the point of halting to snap
him up, but never quite does it,— and so, between
disappointment and expectation, is soon disgusted,
and returns to pursue his more legitimate means of
subsistence.

In striking contrast to this serio-comic strife of
the sparrow and the moth, is the pigeon hawk's
pursuit of the sparrow or the goldfinch. It is a race
of surprising speed and agility. It is a test of wing
and wind. Every muscle is taxed, and every nerve
strained. Such cries of terror and consternation
on the part of the bird, tacking to the right and
left, and making the most desperate efforts to escape, and such silent determination on the part of
the hawk, pressing the bird so closely, flashing and
turning, and timing his movements with those of
the pursued as accurately and as inexorably as if the
two constituted one body, excite feelings of the
deepest concern. You mount the fence or rush out
of your way to see the issue. The only salvation for
the bird is to adopt the tactics of the moth, seeking instantly the cover of some tree, bush, or hedge,
where its smaller size enables it to move about more
rapidly. These pirates are aware of this, and there-

fore prefer to take their prey by one fell swoop.
You may see one of them prowling through an
orchard, with the yellowbirds hovering about him,
crying, *Pi-ty, pi-ty*, in the most desponding tone ;
yet he seems not to regard them, knowing, as do
they, that in the close branches they are as safe
as if in a wall of adamant.

August is the month of the high-sailing hawks.
The hen-hawk is the most noticeable. He likes
the haze and calm of these long, warm days. He
is a bird of leisure, and seems always at his ease.
How beautiful and majestic are his movements!
So self-poised and easy, such an entire absence of
haste, such a magnificent amplitude of circles and
spirals, such a haughty, imperial grace, and, occasionally, such daring aerial evolutions!

With slow, leisurely movement, rarely vibrating
his pinions, he mounts and mounts in an ascending
spiral till he appears a mere speck against the summer sky; then, if the mood seizes him, with wings
half-closed, like a bent bow, he will cleave the air
almost perpendicularly, as if intent on dashing
himself to pieces against the earth; but on nearing
the ground he suddenly mounts again on broad, expanded wing, as if rebounding upon the air, and sails
leisurely away. It is the sublimest feat of the season.
One holds his breath till he sees him rise again.

If inclined to a more gradual and less precipitous descent, he fixes his eye on some distant point

in the earth beneath him, and thither bends his course. He is still almost meteoric in his speed and boldness. You see his path down the heavens, straight as a line; if near, you hear the rush of his wings; his shadow hurtles across the fields, and in an instant you see him quietly perched upon some low tree or decayed stub in a swamp or meadow, with reminiscences of frogs and mice stirring in his maw.

When the south wind blows, it is a study to see three or four of these air-kings at the head of the valley far up toward the mountain, balancing and oscillating upon the strong current; now quite stationary, except a slight tremulous motion like the poise of a rope-dancer, then rising and falling in long undulations, and seeming to resign themselves passively to the wind ; or, again, sailing high and level far above the mountain's peak, no bluster and haste, but, as stated, occasionally a terrible earnestness and speed. Fire at one as he sails overhead, and, unless wounded badly, he will not change his course or gait.

His flight is a perfect picture of repose in motion. It strikes the eye as more surprising than the flight of the pigeon and swallow even, in that the effort put forth is so uniform and delicate as to escape observation, giving to the movement an air of buoyancy and perpetuity, the effluence of power rather than the conscious application of it.

THE RETURN OF THE BIRDS

The calmness and dignity of this hawk, when attacked by crows or the kingbird, are well worthy of him. He seldom deigns to notice his noisy and furious antagonists, but deliberately wheels about in that aerial spiral, and mounts and mounts till his pursuers grow dizzy and return to earth again. It is quite original, this mode of getting rid of an unworthy opponent, rising to heights where the braggart is dazed and bewildered and loses his reckoning! I am not sure but it is worthy of imitation.

But summer wanes, and autumn approaches. The songsters of the seed-time are silent at the reaping of the harvest. Other minstrels take up the strain. It is the heyday of insect life. The day is canopied with musical sound. All the songs of the spring and summer appear to be floating, softened and refined, in the upper air. The birds, in a new but less holiday suit, turn their faces southward. The swallows flock and go ; the bobolinks flock and go; silently and unobserved, the thrushes go. Autumn arrives, bringing finches, warblers, sparrows, and kinglets from the north. Silently the procession passes. Yonder hawk, sailing peacefully away till he is lost in the horizon, is a symbol of the closing season and the departing birds.

BIRCH BROWSINGS

Exploring the woods around Slabsides

EDITOR'S NOTE

In 1860 Burroughs went on a summer camping trip with his friend Elijah Allen, who later introduced him to Whitman. Although Burroughs had been born and raised in the Catskills, this trek into the wilderness of southern New York when he was twenty-three was his first camping trip. Eight years later he joined his brother Hiram and a friend, Hiram Corbin, for a second trip into this same wild region not far from New York City. Burroughs found this summer adventure of 1868 "fruitful to me in much besides trout," and he immediately intended "to make a piece of it—'Among the Birches.'" Subsequently entitled "Birch Browsings," the essay follows a narrative form and a journey motif that make it one of Burroughs's best structured early essays. Primarily a comic account of campers lost in the woods, frying pans spilled in the fire, swarming mosquitoes, and lumpy bedrolls, the narrative humorously dramatizes "how poor an Indian" Burroughs would make and "what a ridiculous figure a party of men may cut in the woods." This lighthearted essay becomes more than a series

EDITOR'S NOTE

of misadventures as the campers fall under the spell of the magical woods. Unlike hunters who have depleted the wild pigeons and deer in the woods, and unlike those who think they have somehow conquered a mountain by climbing to its summit, Burroughs and his companions discover, even then, "one is by no means master of the situation." Written in 1868, "Birch Browsings" first appeared in the *Atlantic* in July 1869 and was reprinted in *Wake-Robin* in 1871.

F.B.

BIRCH BROWSINGS

THE region of which I am about to speak lies in the southern part of the State of New York, and comprises parts of three counties, — Ulster, Sullivan, and Delaware. It is drained by tributaries of both the Hudson and Delaware, and, next to the Adirondack section, contains more wild land than any other tract in the State. The mountains which traverse it, and impart to it its severe northern climate, belong properly to the Catskill range. On some maps of the State they are called the Pine Mountains, though with obvious local impropriety, as pine, so far as I have observed, is nowhere found upon them. "Birch Mountains" would be a more characteristic name, as on their summits birch is the prevailing tree. They are the natural home of the black and yellow birch, which grow here to unusual size. On their sides beech and maple abound; while, mantling their lower slopes and darkening the valleys, hemlock formerly enticed the lumberman and tanner. Except in remote or inaccessible localities, the latter tree is now almost never found. In Shandaken and along the Esopus it is about the only product the country yielded, or is likely to yield. Tanneries by the score have arisen and flourished upon the bark, and some of them still remain. Passing through that region the present season, I saw that the few patches of hemlock that still lingered high up on the sides of the mountains were being felled and peeled, the fresh white boles of the trees, just stripped of their bark, being visible a long distance.

Among these mountains there are no sharp peaks, or abrupt declivities, as in a volcanic region, but long, uniform ranges, heavily timbered to their summits, and delighting the eye with vast, undulating horizon lines. Looking south from the heights about the head of the Delaware, one sees, twenty miles away, a continual succession of blue ranges, one behind the other. If a few large trees are missing on the sky line, one can see the break a long distance off.

Approaching this region from the Hudson River side, you cross a rough, rolling stretch of country, skirting the base of the Catskills, which from a point near Saugerties sweep inland; after a drive of a few hours you are within the shadow of a high, bold mountain, which forms a sort of butt-end to this part of the range, and which is simply called High Point. To the east and southeast it slopes down rapidly to the plain, and looks defiance toward the Hudson, twenty miles distant; in the rear of it, and radiating from it west and northwest, are numerous smaller ranges, backing up, as it were, this haughty chief.

From this point through to Pennsylvania, a distance of nearly one hundred miles, stretches the tract of which I speak. It is a belt of country from twenty to thirty miles wide, bleak and wild, and but sparsely settled. The traveler on the New York and Erie Railroad gets a glimpse of it.

Many cold, rapid trout streams, which flow to all points of the compass, have their source in the small lakes and copious mountain springs of this region. The names of some of them are Mill Brook, Dry Brook, Willewemack, Beaver Kill, Elk Bush Kill, Panther Kill, Neversink, Big Ingin, and Callikoon. Beaver Kill is the main outlet on the west. It joins the Delaware in the wilds of Hancock. The Neversink lays open the region to the south, and also joins the Delaware. To the east, various Kills unite with the Big Ingin to form the Esopus, which flows into the Hudson. Dry Brook and Mill Brook, both famous trout streams, from twelve to fifteen miles long, find their way into the Delaware.

The east or Pepacton branch of the Delaware itself takes its rise near here in a deep pass between the mountains. I have many times drunk at a copious spring by the roadside, where the infant river first sees the light. A few yards beyond, the water flows the other way, directing its course through the Bear Kill and Schoharie Kill into the Mohawk.

Such game and wild animals as still linger in the State are found in this region. Bears occasionally make havoc among the sheep. The clearings at the head of a valley are oftenest the scene of their depredations.

Wild pigeons, in immense numbers, used to breed regularly in the valley of the Big Ingin and about the head of the Neversink. The treetops for miles were full of their nests, while the going and coming of the old birds kept up a constant din. But the gunners soon got wind of it, and from far and near were wont to pour in during the spring, and to slaughter both old and young. This practice soon had the effect of driving the pigeons all away, and now only a few pairs breed in these woods.

Deer are still met with, though they are becoming scarcer every year. Last winter near seventy head were killed on the Beaver Kill alone. I heard of one wretch, who, finding the deer snowbound, walked up to them on his snowshoes, and one morning before breakfast slaughtered six, leaving their carcasses where they fell. There are traditions of persons having been smitten blind or senseless when about to commit some heinous offense, but the fact that this villain escaped without some such visitation throws discredit on all such stories.

The great attraction, however, of this region, is the brook trout, with which the streams and lakes abound. The water is of excessive coldness, the thermometer indicating 44° and 45° in the springs, and 47° or 48° in the smaller streams. The trout are generally small, but in the more remote branches their number is very great. In such localities the fish are quite black, but in the lakes they are of a lustre and brilliancy impossible to describe.

These waters have been much visited of late years by fishing parties, and the name of Beaver Kill is now a potent word among New York sportsmen.

One lake, in the wilds of Callikoon, abounds in a peculiar species of white sucker, which is of excellent quality. It is taken only in spring, during the spawning season, at the time "when the leaves are as big as a chipmunk's ears." The fish run up the small streams and inlets, beginning at nightfall, and continuing till the channel is literally packed with them, and every inch of space is occupied. The fishermen pounce upon them at such times, and scoop them up by the bushel, usually wading right into the living mass and landing the fish with their hands. A small party will often secure in this manner a wagon-load of fish. Certain conditions of the weather, as a warm south or southwest wind, are considered most favorable for the fish to run.

Though familiar all my life with the outskirts of this region, I have only twice dipped into its wilder portions. Once in 1860 a friend and myself traced the Beaver Kill to its source, and encamped by Balsam Lake. A cold and protracted rainstorm coming on, we were obliged to leave the woods before we were ready. Neither of us will soon forget that tramp by an unknown route over the mountains, encumbered as we were with a hundred and one superfluities which we had foolishly brought along to solace ourselves with in the woods ; nor that halt on the summit, where we cooked and ate our fish in a drizzling rain; nor, again, that rude log house, with its sweet hospitality, which we reached just at nightfall on Mill Brook.

In 1868 a party of three of us set out for a brief trouting excursion to a body of water called Thomas's Lake, situated in the same chain of mountains. On this excursion, more particularly than on any other I have ever undertaken, I was taught how poor an Indian I should make, and what a ridiculous figure a party of men may cut in the woods when the way is uncertain and the mountains high.

We left our team at a farmhouse near the head of the Mill Brook, one June afternoon, and with knapsacks on our shoulders struck into the woods at the base of the mountain, hoping to cross the range that intervened between us and the lake by sunset. We engaged a good-natured but rather indolent young man, who happened to be stopping at the house, and who had carried a knapsack in the Union armies, to pilot us a couple of miles into the woods so as to guard against any mistakes at the outset. It seemed the easiest thing in the world to find the lake. The lay of the land was so simple, according to accounts, that I felt sure I could go to it in the dark. "Go up this little brook to its source on the side of the mountain," they said. "The valley that contains the lake heads directly on the other side." What could be easier! But on a little further inquiry, they said we should "bear well to the left" when we reached the top of the mountain. This opened the doors again; "bearing well to the left" was an uncertain performance in strange woods. We might bear so well to the left that it would bring us ill. But why bear to the left at all, if the lake was directly opposite? Well, not quite opposite; a little to the left. There were two or three other valleys that headed in near there. We could easily find the right one. But to make assurance doubly sure, we engaged a guide, as stated, to give us a good start, and go with us beyond the bearing-to-the-left point. He had been to the lake the winter before and knew the way. Our course, the first half hour, was along an obscure wood-road which had been used for drawing ash logs off the mountain in winter. There was some hemlock, but more maple and birch. The woods were dense and free from underbrush, the ascent gradual. Most of the way we kept the voice of the creek in our ear on the right. I approached it once, and found it swarming with trout. The water was as cold as one ever need wish. After a while the ascent grew steeper, the creek became a mere rill that issued from beneath loose, moss-covered rocks and stones, and with much labor and puffing we drew ourselves up the rugged declivity. Every mountain has its steepest point, which is usually near the summit, in keeping, I suppose, with the providence that makes the darkest hour just before day. It is steep, steeper, steepest, till you emerge on the smooth level or gently rounded space at the top, which the old ice-gods polished off so long ago.

We found this mountain had a hollow in its back where the ground was soft and swampy. Some gigantic ferns, which we passed through, came nearly to our shoulders. We passed also several patches of swamp honeysuckles, red with blossoms.

Our guide at length paused on a big rock where the land began to dip down the other way, and concluded that he had gone far enough, and that we would now have no difficulty in finding the lake. "It must lie right down there," he said, pointing with his hand. But it was plain that he was not quite sure in his own mind. He had several times wavered in his course, and had shown considerable embarrassment when bearing to the left across the summit. Still we thought little of it. We were full of confidence, and, bidding him adieu, plunged down the mountain-side, following a spring run that we had no doubt led to the lake.

In these woods, which had a southeastern exposure, I first began to notice the wood thrush. In coming up the other side I had not seen a feather of any kind, or heard a note. Now the golden *trillide-de* of the wood thrush sounded through the silent woods. While looking for a fish-pole about halfway down the mountain, I saw a thrush's nest in a little sapling about ten feet from the ground.

After continuing our descent till our only guide, the spring run, became quite a trout brook, and its tiny murmur a loud brawl, we began to peer anxiously through the trees for a glimpse of the lake, or for some conformation of the land that would indicate its proximity. An object which we vaguely discerned in looking under the near trees and over the more distant ones proved, on further inspection, to be a patch of plowed ground. Presently we made out a burnt fallow near it. This was a wet blanket to our enthusiasm. No lake, no sport, no trout for supper that night. The rather indolent young man had either played us a trick, or, as seemed more likely, had missed the way. We were particularly anxious to be at the lake between sundown and dark, as at that time the trout jump most freely.

Pushing on, we soon emerged into a stumpy field, at the head of a steep valley, which swept around toward the west. About two hundred rods below us was a rude log house, with smoke issuing from the chimney. A boy came out and moved toward the spring with a pail in his hand. We shouted to him, when he turned and ran back into the house without pausing to reply. In a moment the whole family hastily rushed into the yard, and turned their faces toward us. If we had come down their chimney, they could not have seemed more aston- ished. Not making out what they said, I went down to the house, and learned to my chagrin that we were still on the Mill Brook side, having crossed only a spur of the mountain. We had not borne sufficiently to the left, so that the main range, which, at the point of crossing, suddenly breaks off to the southeast, still intervened between us and the lake. We were about five miles, as the water runs, from the point of starting, and over two from the lake. We must go directly back to the top of the range where the guide had left us, and then, by keeping well to the left, we would soon come to a line of marked trees, which would lead us to the lake. So, turning upon our trail, we doggedly began the work of undoing what we had just done, — in all cases a disagreeable task, in this case a very laborious one also. It was after sunset when we turned back, and before we had got halfway up the mountain, it began to be quite dark. We were often obliged to rest our packs against trees and take breath, which made our progress slow. Finally a halt was called, beside an immense flat rock which had paused on its slide down the mountain, and we prepared to encamp for the night. A fire was built, the rock cleared off, a small ration of bread served out, our accoutrements hung up out of the way of the hedgehogs that were supposed to infest the locality, and then we disposed ourselves for sleep. If the owls or porcupines (and I think I heard one of the latter in the middle of the night) reconnoitred our camp, they saw a buffalo robe spread upon a rock, with three old felt hats arranged on one side, and three pairs of sorry-looking cowhide boots protruding from the other.

When we lay down, there was apparently not a mosquito in the woods; but the "no-see-ems," as Thoreau's Indian aptly named the midges, soon found us out, and after the fire had gone down, annoyed us much. My hands and wrists suddenly began to smart and itch in a most unaccountable manner. My first thought was that they had been poisoned in some way. Then the smarting extended to my neck and face, even to my scalp, when I began to suspect what was the matter. So, wrapping myself up more thoroughly, and stowing my hands away as best I could, I tried to sleep, being some time behind my companions, who appeared not to mind the "no-see-ems." I was further annoyed by some little irregularity on my side of the couch. The chambermaid had not beaten it up well. One huge lump refused to be mollified, and each attempt to adapt it to some natural hollow in my own body brought only a moment's relief. But at last I got the better of this also and slept. Late in the night I woke up, just in time to hear a golden-crowned thrush sing in a tree near by. It sang as loud and cheerily as at midday, and I thought myself, after all, quite in luck. Birds occasionally sing at night, just as the cock crows. I have heard the hairbird, and the note of the king-bird; and the ruffed grouse frequently drums at night.

At the first faint signs of day a wood thrush sang, a few rods below us. Then after a little delay, as the gray light began to grow around, thrushes broke out in full song in all parts of the woods. I thought I had never before heard them sing so sweetly. Such a leisurely, golden chant! — it consoled us for all we had undergone. It was the first thing in order, — the worms were safe till after this morning chorus. I judged that the birds roosted but a few feet from the ground. In fact, a bird in all cases roosts where it builds, and the wood thrush occupies, as it were, the first story of the woods.

There is something singular about the distribution of the wood thrushes. At an earlier stage of my observations I should have been much surprised at finding it in these woods. Indeed, I had stated

in print on two occasions that the wood thrush was
not found in the higher lands of the Catskills, but
that the hermit thrush and the veery, or Wilson's
thrush, were common. It turns out that the statement is only half true. The wood thrush is found
also, but is much more rare and secluded in its habits than either of the others, being seen only during
the breeding season on remote mountains, and then
only on their eastern and southern slopes. I have
never yet in this region found the bird spending
the season in the near and familiar woods, which is
directly contrary to observations I have made in
other parts of the State. So different are the habits of birds in different localities.

As soon as it was fairly light we were up and
ready to resume our march. A small bit of bread
and butter and a swallow or two of whiskey was all
we had for breakfast that morning. Our supply of
each was very limited, and we were anxious to save
a little of both, to relieve the diet of trout to which
we looked forward.

At an early hour we reached the rock where we
had parted with the guide, and looked around us
into the dense, trackless woods with many misgivings. To strike out now on our own hook, where
the way was so blind and after the experience we
had just had, was a step not to be carelessly taken.
The tops of these mountains are so broad, and a
short distance in the woods seems so far, that one

is by no means master of the situation after reaching the summit. And then there are so many spurs and offshoots and changes of direction, added to the impossibility of making any generalization by the aid of the eye, that before one is aware of it he is very wide of his mark.

I remembered now that a young farmer of my acquaintance had told me how he had made a long day's march through the heart of this region, without path or guide of any kind, and had hit his mark squarely. He had been barkpeeling in Callikoon, — a famous country for bark, — and, having got enough of it, he desired to reach his home on Dry Brook without making the usual circuitous journey between the two places. To do this necessitated a march of ten or twelve miles across several ranges of mountains and through an unbroken forest, — a hazardous undertaking in which no one would join him. Even the old hunters who were familiar with the ground dissuaded him and predicted the failure of his enterprise. But having made up his mind, he possessed himself thoroughly of the topography of the country from the aforesaid hunters, shouldered his axe, and set out, holding a straight course through the woods, and turning aside for neither swamps, streams, nor mountains. When he paused to rest he would mark some object ahead of him with his eye, in order that on getting up again he might not deviate from his course. His

directors had told him of a hunter's cabin about
midway on his route, which if he struck he might
be sure he was right. About noon this cabin was
reached, and at sunset he emerged at the head of
Dry Brook.

After looking in vain for the line of marked
trees, we moved off to the left in a doubtful, hesi-
tating manner, keeping on the highest ground and
blazing the trees as we went. We were afraid to
go downhill, lest we should descend too soon; our
vantage-ground was high ground. A thick fog coming on, we were more bewildered than ever. Still
we pressed forward, climbing up ledges and wading
through ferns for about two hours, when we paused
by a spring that issued from beneath an immense
wall of rock that belted the highest part of the
mountain. There was quite a broad plateau here,
and the birch wood was very dense, and the trees
of unusual size.

After resting and exchanging opinions, we all
concluded that it was best not to continue our search
encumbered as we were; but we were not willing to
abandon it altogether, and I proposed to my companions to leave them beside the spring with our
traps, while I made one thorough and final effort to
find the lake. If I succeeded and desired them to
come forward, I was to fire my gun three times; if
I failed and wished to return, I would fire it twice,
they of course responding.

So, filling my canteen from the spring, I set out again, taking the spring run for my guide. Before I had followed it two hundred yards, it sank into the ground at my feet. I had half a mind to be superstitious and to believe that we were under a spell, since our guides played us such tricks. However, I determined to put the matter to a further test, and struck out boldly to the left. This seemed to be the keyword, — to the left, to the left. The fog had now lifted, so that I could form a better idea of the lay of the land. Twice I looked down the steep sides of the mountain, sorely tempted to risk a plunge. Still I hesitated and kept along on the brink. As I stood on a rock deliberating, I heard a crackling of the brush, like the tread of some large game, on a plateau below me. Suspecting the truth of the case, I moved stealthily down, and found a herd of young cattle leisurely browsing. We had several times crossed their trail, and had seen that morning a level, grassy place on the top of the mountain, where they had passed the night. Instead of being frightened, as I had expected, they seemed greatly delighted, and gathered around me as if to inquire the tidings from the outer world, — perhaps the quotations of the cattle market. They came up to me, and eagerly licked my hand, clothes, and gun. Salt was what they were after, and they were ready to swallow anything that contained the smallest percentage of it. They were mostly yearlings and as sleek as moles. They had a very gamy look. We were afterwards told that, in the spring, the farmers round about turn into these woods their young cattle, which do not come out again till fall. They are then in good condition, — not fat, like grass-fed cattle, but trim and supple, like deer. Once a month the owner hunts them up and salts them. They have their beats, and seldom wander beyond well-defined limits. It was interesting to see them feed. They browsed on the low limbs and bushes, and on the various plants, munching at everything without any apparent discrimination.

They attempted to follow me, but I escaped them by clambering down some steep rocks. I now found myself gradually edging down the side of the mountain, keeping around it in a spiral manner, and scanning the woods and the shape of the ground for some encouraging hint or sign. Finally the woods became more open, and the descent less rapid. The trees were remarkably straight and uniform in size. Black birches, the first I had seen, were very numerous. I felt encouraged. Listening attentively, I caught, from a breeze just lifting the drooping leaves, a sound that I willingly believed was made by a bullfrog. On this hint, I tore down through the woods at my highest speed. Then I paused and listened again. This time there was no mistaking it; it was the sound of frogs. Much elated, I rushed on. By and by I could hear them as I ran. *Pthrung, pthrung,* croaked the old ones; *pug, pug,* shrilly joined in the smaller fry.

Then I caught, through the lower trees, a gleam of blue, which I first thought was distant sky. A second look and I knew it to be water, and in a moment more I stepped from the woods and stood upon the shore of the lake. I exulted silently. There it was at last, sparkling in the morning sun, and as beautiful as a dream. It was so good to come upon such open space and such bright hues, after wandering in the dim, dense woods! The eye is as delighted as an escaped bird, and darts gleefully from point to point.

The lake was a long oval, scarcely more than a mile in circumference, with evenly wooded shores, which rose gradually on all sides. After contemplating the scene for a moment, I stepped back into the woods, and, loading my gun as heavily as I dared, discharged it three times. The reports seemed to fill all the mountains with sound. The frogs quickly hushed, and I listened for the response. But no response came. Then I tried again and again, but without evoking an answer. One of my companions, however, who had climbed to the top of the high rocks in the rear of the spring, thought he heard faintly one report. It seemed an immense distance below him, and far around under the mountain. I knew I had come a long way, and hardly expected to be able to communicate with

my companions in the manner agreed upon. I
therefore started back, choosing my course without
any reference to the circuitous route by which I had
come, and loading heavily and firing at intervals.
I must have aroused many long-dormant echoes
from a Rip Van Winkle sleep. As my powder got
low, I fired and halloed alternately, till I came near
splitting both my throat and gun. Finally, after I
had begun to have a very ugly feeling of alarm and
disappointment, and to cast about vaguely for some
course to pursue in the emergency that seemed near
at hand, — namely, the loss of my companions now
I had found the lake, — a favoring breeze brought
me the last echo of a response. I rejoined with
spirit, and hastened with all speed in the direction
whence the sound had come, but, after repeated trials, failed to elicit another answering sound. This
filled me with apprehension again. I feared that
my friends had been misled by the reverberations,
and I pictured them to myself hastening in the
opposite direction. Paying little attention to my
course, but paying dearly for my carelessness afterward, I rushed forward to undeceive them. But
they had not been deceived, and in a few moments an answering shout revealed them near at
hand. I heard their tramp, the bushes parted, and
we three met again.

In answer to their eager inquiries, I assured them
that I had seen the lake, that it was at the foot of

the mountain, and that we could not miss it if we kept straight down from where we then were.

My clothes were soaked with perspiration, but I shouldered my knapsack with alacrity, and we began the descent. I noticed that the woods were much thicker, and had quite a different look from those I had passed through, but thought nothing of it, as I expected to strike the lake near its head, whereas I had before come out at its foot. We had not gone far when we crossed a line of marked trees, which my companions were disposed to follow. It intersected our course nearly at right angles, and kept along and up the side of the mountain. My impression was that it led up from the lake, and that by keeping our own course we should reach the lake sooner than if we followed this line.

About halfway down the mountain, we could see through the interstices the opposite slope. I encouraged my comrades by telling them that the lake was between us and that, and not more than half a mile distant. We soon reached the bottom, where we found a small stream and quite an extensive alder swamp, evidently the ancient bed of a lake. I explained to my half-vexed and half-incredulous companions that we were probably above the lake, and that this stream must lead to it. "Follow it," they said; "we will wait here till we hear from you."

So I went on, more than ever disposed to believe

that we were under a spell, and that the lake had slipped from my grasp after all. Seeing no favorable sign as I went forward, I laid down my accoutrements, and climbed a decayed beech that leaned out over the swamp and promised a good view from the top. As I stretched myself up to look around from the highest attainable branch, there was suddenly a loud crack at the root. With a celerity that would at least have done credit to a bear, I regained the ground, having caught but a momentary glimpse of the country, but enough to convince me no lake was near. Leaving all incumbrances here but my gun, I still pressed on, loath to be thus baffled. After floundering through another alder swamp for nearly half a mile, I flattered myself that I was close on to the lake. I caught sight of a low spur of the mountain sweeping around like a half-extended arm, and I fondly imagined that within its clasp was the object of my search. But I found only more alder swamp. After this region was cleared, the creek began to descend the mountain very rapidly. Its banks became high and narrow, and it went whirling away with a sound that seemed to my ears like a burst of ironical laughter. I turned back with a feeling of mingled disgust, shame, and vexation. In fact I was almost sick, and when I reached my companions, after an absence of nearly two hours, hungry, fatigued, and disheartened, I would have sold my interest in

Thomas's Lake at a very low figure. For the first time, I heartily wished myself well out of the woods. Thomas might keep his lake, and the enchanters guard his possession! I doubted if he had ever found it the second time, or if any one else ever had.

My companions, who were quite fresh, and who had not felt the strain of baffled purpose as I had, assumed a more encouraging tone. After I had rested awhile, and partaken sparingly of the bread and whiskey, which in such an emergency is a great improvement on bread and water, I agreed to their proposition that we should make another attempt. As if to reassure us, a robin sounded his cheery call near by, and the winter wren, the first I had heard in these woods, set his music-box going, which fairly ran over with fine, gushing, lyrical sounds. There can be no doubt but this bird is one of our finest songsters. If it would only thrive and sing well when caged, like the canary, how far it would surpass that bird! It has all the vivacity and versatility of the canary, without any of its shrillness. Its song is indeed a little cascade of melody.

We again retraced our steps, rolling the stone, as it were, back up the mountain, determined to commit ourselves to the line of marked trees. These we finally reached, and, after exploring the country to the right, saw that bearing to the left was still the order. The trail led up over a gentle rise of ground, and in less than twenty minutes we were in the woods I had passed through when I found the lake. The error I had made was then plain: we had come off the mountain a few paces too far to the right, and so had passed down on the wrong side of the ridge, into what we afterwards learned was the valley of Alder Creek.

We now made good time, and before many minutes I again saw the mimic sky glance through the trees. As we approached the lake, a solitary woodchuck, the first wild animal we had seen since entering the woods, sat crouched upon the root of a tree a few feet from the water, apparently completely nonplused by the unexpected appearance of danger on the land side. All retreat was cut off, and he looked his fate in the face without flinching. I slaughtered him just as a savage would have done, and from the same motive, — I wanted his carcass to eat.

The mid-afternoon sun was now shining upon the lake, and a low, steady breeze drove the little waves rocking to the shore. A herd of cattle were browsing on the other side, and the bell of the leader sounded across the water. In these solitudes its clang was wild and musical.

To try the trout was the first thing in order. On a rude raft of logs which we found moored at the shore, and which with two aboard shipped about a foot of water, we floated out and wet our first fly in Thomas's Lake; but the trout refused to jump, and, to be frank, not more than a dozen and a half were caught during our stay. Only a week previous, a party of three had taken in a few hours all the fish they could carry out of the woods, and had nearly surfeited their neighbors with trout. But from some cause they now refused to rise, or to touch any kind of bait: so we fell to catching the sunfish, which were small but very abundant. Their nests were all along shore. A space about the size of a breakfast-plate was cleared of sediment and decayed vegetable matter, revealing the pebbly bottom, fresh and bright, with one or two fish suspended over the centre of it, keeping watch and ward. If an intruder approached, they would dart at him spitefully. These fish have the air of bantam cocks, and, with their sharp, prickly fins and spines and scaly sides, must be ugly customers in a hand-to-hand encounter with other finny warriors. To a hungry man they look about as unpromising as hemlock slivers, so thorny and thin are they; yet there is sweet meat in them, as we found that day.

Much refreshed, I set out with the sun low in the west to explore the outlet of the lake and try for trout there, while my companions made further trials in the lake itself. The outlet, as is usual in bodies of water of this kind, was very gentle and private. The stream, six or eight feet wide, flowed silently and evenly along for a distance of three or four rods, when it suddenly, as if conscious of its freedom, took a leap down some rocks. Thence, as far as I followed it, its descent was very rapid through a continuous succession of brief falls like so many steps down the mountain. Its appearance promised more trout than I found, though I returned to camp with a very respectable string.

Toward sunset I went round to explore the inlet, and found that as usual the stream wound leisurely through marshy ground. The water being much colder than in the outlet, the trout were more plentiful. As I was picking my way over the miry ground and through the rank growths, a ruffed grouse hopped up on a fallen branch a few paces before me, and, jerking his tail, threatened to take flight. But as I was at that moment gunless and remained stationary, he presently jumped down and walked away.

A seeker of birds, and ever on the alert for some new acquaintance, my attention was arrested, on first entering the swamp, by a bright, lively song, or warble, that issued from the branches overhead, and that was entirely new to me, though there was something in the tone of it that told me the bird was related to the wood-wagtail and to the water-wagtail or thrush. The strain was emphatic and quite loud, like the canary's, but very brief. The bird kept itself well secreted in the upper branches of the trees, and for a long time eluded my eye. I passed to and fro several times, and it seemed to break out afresh as I approached a certain little bend in the creek, and to cease after I had got beyond it; no doubt its nest was somewhere in the vicinity. After some delay the bird was sighted and brought down. It proved to be the small, or northern, water-thrush (called also the New York water-thrush), — a new bird to me. In size it was noticeably smaller than the large, or Louisiana, water-thrush, as described by Audubon, but in other respects its general appearance was the same. It was a great treat to me, and again I felt myself in luck.

This bird was unknown to the older ornithologists, and is but poorly described by the new. It builds a mossy nest on the ground, or under the edge of a decayed log. A correspondent writes me that he has found it breeding on the mountains in Pennsylvania. The large-billed water-thrush is much the superior songster, but the present species has a very bright and cheerful strain. The specimen I saw, contrary to the habits of the family, kept in the treetops like a warbler, and seemed to be engaged in catching insects.

The birds were unusually plentiful and noisy about the head of this lake; robins, blue jays, and woodpeckers greeted me with their familiar notes. The blue jays found an owl or some wild animal a short distance above me, and, as is their custom on such occasions, proclaimed it at the top of their voices, and kept on till the darkness began to gather in the woods.

I also heard here, as I had at two or three other points in the course of the day, the peculiar, resonant hammering of some species of woodpecker upon the hard, dry limbs. It was unlike any sound of the kind I had ever before heard, and, repeated at intervals through the silent woods, was a very marked and characteristic feature. Its peculiarity was the ordered succession of the raps, which gave it the character of a premeditated performance. There were first three strokes following each other rapidly, then two much louder ones with longer intervals between them. I heard the drumming here, and the next day at sunset at Furlow Lake, the source of Dry Brook, and in no instance was the order varied. There was melody in it, such as a woodpecker knows how to evoke from a smooth, dry branch. It suggested something quite as pleasing as the liveliest bird-song, and was if anything more woodsy and wild. As the yellow-bellied woodpecker was the most abundant species in these woods, I attributed it to him. It is the one sound that still links itself with those scenes in my mind.

At sunset the grouse began to drum in all parts of the woods about the lake. I could hear five at one time, *thump, thump, thump, thump, thr-r-r-r-r-r-rr*. It was a homely, welcome sound. As I returned to camp at twilight, along the shore of the lake, the frogs also were in full chorus. The older ones ripped out their responses to each other with terrific force and volume. I know of no other animal capable of giving forth so much sound, in proportion to its size, as a frog. Some of these seemed to bellow as loud as a two-year-old bull. They were of immense size, and very abundant. No frog-eater had ever been there. Near the shore we felled a tree which reached far out in the lake. Upon the trunk and branches the frogs soon collected in large numbers, and gamboled and splashed about the half-submerged top, like a parcel of schoolboys, making nearly as much noise.

After dark, as I was frying the fish, a panful of the largest trout was accidently capsized in the fire. With rueful countenances we contemplated the irreparable loss our commissariat had sustained by this mishap; but remembering there was virtue in ashes, we poked the half-consumed fish from the bed of coals and ate them, and they were good.

We lodged that night on a brush-heap and slept soundly. The green, yielding beech-twigs, covered with a buffalo robe, were equal to a hair mattress. The heat and smoke from a large fire kindled in the afternoon had banished every "no-see-em" from the locality, and in the morning the sun was above the mountain before we awoke.

I immediately started again for the inlet, and went far up the stream toward its source. A fair

string of trout for breakfast was my reward. The
cattle with the bell were at the head of the valley,
where they had passed the night. Most of them
were two-year-old steers. They came up to me and
begged for salt, and scared the fish by their impor-
tunities.

We finished our bread that morning, and ate
every fish we could catch, and about ten o'clock
prepared to leave the lake. The weather had been
admirable, and the lake was a gem, and I would
gladly have spent a week in the neighborhood; but
the question of supplies was a serious one, and
would brook no delay.

When we reached, on our return, the point where
we had crossed the line of marked trees the day
before, the question arose whether we should still
trust ourselves to this line, or follow our own trail
back to the spring and the battlement of rocks on
the top of the mountain, and thence to the rock
where the guide had left us. We decided in favor
of the former course. After a march of three quarters of an hour the blazed trees ceased, and we
concluded we were near the point at which we had
parted with the guide. So we built a fire, laid
down our loads, and cast about on all sides for some
clew as to our exact locality. Nearly an hour was
consumed in this manner, and without any result.
I came upon a brood of young grouse, which di-
verted me for a moment. The old one blustered

about at a furious rate, trying to draw all attention to herself, while the young ones, which were unable to fly, hid themselves. She whined like a dog in great distress, and dragged herself along apparently with the greatest difficulty. As I pursued her, she ran very nimbly, and presently flew a few yards. Then, as I went on, she flew farther and farther each time, till at last she got up, and went humming through the woods as if she had no interest in them. I went back and caught one of the young, which had simply squatted close to the leaves. I took it up and set it on the palm of my hand, which it hugged as closely as if still upon the ground. I then put it in my coatsleeve, when it ran and nestled in my armpit.

When we met at the sign of the smoke, opinions differed as to the most feasible course. There was no doubt but that we could get out of the woods; but we wished to get out speedily, and as near as possible to the point where we had entered. Half ashamed of our timidity and indecision, we finally tramped away back to where we had crossed the line of blazed trees, followed our old trail to the spring on the top of the range, and, after much searching and scouring to the right and left, found ourselves at the very place we had left two hours before. Another deliberation and a divided council. But something must be done. It was then mid-afternoon, and the prospect of spending another night on the mountains, without food or drink, was not pleasant. So we moved down the ridge. Here another line of marked trees was found, the course of which formed an obtuse angle with the one we had followed. It kept on the top of the ridge for perhaps a mile, when it entirely disappeared, and we were as much adrift as ever. Then one of the party swore an oath, and said he was going out of those woods, hit or miss, and, wheeling to the right, instantly plunged over the brink of the mountain. The rest followed, but would fain have paused and ciphered away at their own uncertainties, to see if a certainty could not be arrived at as to where we would come out. But our bold leader was solving the problem in the right way. Down and down and still down we went, as if we were to bring up in the bowels of the earth. It was by far the steepest descent we had made, and we felt a grim satisfaction in knowing that we could not retrace our steps this time, be the issue what it might. As we paused on the brink of a ledge of rocks, we chanced to see through the trees distant cleared land. A house or barn also was dimly descried. This was encouraging ; but we could not make out whether it was on Beaver Kill or Mill Brook or Dry Brook, and did not long stop to consider where it was. We at last brought up at the bottom of a deep gorge, through which flowed a rapid creek that literally swarmed with trout. But we were in no mood to catch them, and pushed on along the channel of the stream, sometimes leaping from rock to rock, and sometimes splashing heedlessly through the water, and speculating the while as to where we should probably come out. On the Beaver Kill, my companions thought; but, from the position of the sun, I said, on the Mill Brook, about six miles below our team; for I remembered having seen, in coming up this stream, a deep, wild valley that led up into the mountains, like this one. Soon the banks of the stream became lower, and we moved into the woods. Here we entered upon an obscure wood-road, which presently conducted us into the midst of a vast hemlock forest. The land had a gentle slope, and we wondered why the lumbermen and barkmen who prowl through these woods had left this fine tract untouched. Beyond this the forest was mostly birch and maple.

We were now close to the settlement, and began to hear human sounds. One rod more, and we were out of the woods. It took us a moment to comprehend the scene. Things looked very strange at first; but quickly they began to change and to put on familiar features. Some magic scene-shifting seemed to take place before my eyes, till, instead of the unknown settlement which I at first seemed to look upon, there stood the farmhouse at which we had stopped two days before, and at the same moment we heard the stamping of our team in the barn. We sat down and laughed heartily over our good luck. Our desperate venture had resulted better than we had dared to hope, and had shamed our wisest plans. At the house our arrival had been anticipated about this time, and dinner was being put upon the table.

It was then five o'clock, so that we had been in the woods just forty-eight hours; but if time is only phenomenal, as the philosophers say, and life only in feeling, as the poets aver, we were some months, if not years, older at that moment than we had been two days before. Yet younger, too, — though this be a paradox, — for the birches had infused into us some of their own suppleness and strength.

THE FOX

A winter afternoon by the fire at Slabsides

EDITOR'S NOTE

"The Fox" appeared in Burroughs's second nature book, *Winter Sunshine,* published in 1875. Burroughs began work on this book while living in Washington, D.C., and completed it after moving to West Park, New York, where he settled along the Hudson River on the fruit farm he named Riverby. He was thirty-six years old and for the remaining forty-eight years of his life, while living at Riverby, he produced a book on the average of every two years. In *Winter Sunshine,* as the title suggests, he broadened his subjects and themes beyond those of his first nature book which had identified him as an author of essays primarily about spring and summer birds. His second book contained essays about people met on country roads, the national capital in wintertime, winter animals, wild apple trees, and trips to England and France. In expanding "The Fox," first printed in *Putnam's* in October 1870, Burroughs added a personal account of a fox hunt. As with his articles about birds, where he drew on authorities like Audubon, Wilson, Nuttal, and Baird, Burroughs always balanced and

EDITOR'S NOTE

tested secondary sources through personal experience. Through a blending of established fact, amusing anecdote, and personal observation, Burroughs conveys his lifelong fascination and admiration for the fox as both predator and prey. Scientific knowledge alone is not enough to account for an animal whose apparent ability to shrink in size and to survive extinction seemingly defies common sense.

F.B.

THE FOX

I HAVE already spoken of the fox at some length, but it will take a chapter by itself to do half justice to his portrait.

He furnishes, perhaps, the only instance that can be cited of a fur-bearing animal that not only holds its own, but that actually increases in the face of the means that are used for its extermination. The beaver, for instance, was gone before the earliest settlers could get a sight of him; and even the mink and marten are now only rarely seen, or not seen at all, in places where they were once abundant.

But the fox has survived civilization, and in some localities is no doubt more abundant now than in the time of the Revolution. For half a century at least he has been almost the only prize, in the way of fur, that was to be found on our mountains, and he has been hunted and trapped and waylaid, sought for as game and pursued in enmity, taken by fair means and by foul, and yet there seems not the slightest danger of the species becoming extinct.

One would think that a single hound in a neighborhood, filling the mountains with his bayings, and leaving no nook or byway of them unexplored, was enough to drive and scare every fox from the country. But not so. Indeed, I am almost tempted to say, the more hounds, the more foxes.

I recently spent a summer month in a mountainous district in the State of New York, where, from its earliest settlement, the red fox has been the standing prize for skill in the use of the trap and gun. At the house where I was staying were two foxhounds, and a neighbor half a mile distant had a third. There were many others in the township, and in season they were well employed, too; but the three spoken of, attended by their owners, held high carnival on the mountains in the immediate vicinity. And many were the foxes that, winter after winter, fell before them, twenty-five having been shot, the season before my visit, on one small range alone. And yet the foxes were apparently never more abundant than they were that summer, and never bolder, coming at night within a few rods of the house, and of the unchained alert hounds, and making havoc among the poultry.

One morning a large, fat goose was found minus her head and otherwise mangled. Both hounds had disappeared, and, as they did not come back till near night, it was inferred that they had cut short Reynard's repast, and given him a good chase into the bargain. But next night he was back again, and this time got safely off with the goose. A couple of nights after he must have come with recruits, for next morning three large goslings were reported missing. The silly geese now got it through their noddles that there was danger about, and every night thereafter came close up to the house to roost.

A brood of turkeys, the old one tied to a tree a few rods to the rear of the house, were the next objects of attack. The predaceous rascal came, as usual, in the latter half of the night. I happened to be awake, and heard the helpless turkey cry "quit," "quit," with great emphasis. Another sleeper, on the floor above me, who, it seems, had been sleeping with one ear awake for several nights in apprehension for the safety of his turkeys, heard the sound also, and instantly divined its cause. I heard the window open and a voice summon the dogs. A loud bellow was the response, which caused Reynard to take himself off in a hurry. A moment more, and the mother turkey would have shared the fate of the geese. There she lay at the end of her tether, with extended wings, bitten and rumpled. The young ones, roosting in a row on the fence near by, had taken flight on the first alarm.

Turkeys, retaining many of their wild instincts, are less easily captured by the fox than any other of our domestic fowls. On the slightest show of danger they take to wing, and it is not unusual, in the locality of which I speak, to find them in the morning perched in the most unwonted places, as on the peak of the barn or hay-shed, or on the tops of the apple-trees, their tails spread and their manners showing much excitement. Perchance one turkey is minus her tail, the fox having succeeded in getting only a mouthful of quills.

As the brood grows and their wings develop, they wander far from the house in quest of grasshoppers. At such times they are all watchfulness and suspicion. Crossing the fields one day, attended by a dog that much resembled a fox, I came suddenly upon a brood about one third grown, which were feeding in a pasture just beyond a wood. It so happened that they caught sight of the dog without seeing me, when instantly, with the celerity of wild game, they launched into the air, and, while the old one perched upon a treetop, as if to keep an eye on the supposed enemy, the young went sailing over the trees toward home.

The two hounds above referred to, accompanied by a cur-dog, whose business it was to mind the farm, but who took as much delight in running away from prosy duty as if he had been a schoolboy, would frequently steal off and have a good hunt all by themselves, just for the fun of the thing, I suppose. I more than half suspect that it was as a kind of taunt or retaliation, that Reynard came and took the geese from under their very noses. One morning they went off and stayed till the afternoon of the next day; they ran the fox all day and all night, the hounds baying at every jump, the cur-dog silent and tenacious. When the trio returned, they came dragging themselves along, stiff, footsore, gaunt, and hungry. For a day or two afterward they lay about the kennels, seeming to dread nothing so much as the having to move. The stolen hunt was their "spree," their "bender," and of course they must take time to get over it.

Some old hunters think the fox enjoys the chase as much as the hound, especially when the latter runs slow, as the best hounds do. The fox will wait for the hound, will sit down and listen, or play about, crossing and recrossing and doubling upon his track, as if enjoying a mischievous consciousness of the perplexity he would presently cause his pursuer. It is evident, however, that the fox does not always have his share of the fun: before a swift dog, or in a deep snow, or on a wet day, when his tail gets heavy, he must put his best foot forward. As a last resort he "holes up." Sometimes he resorts to numerous devices to mislead and escape the dog altogether. He will walk in the bed of a small creek, or on a rail-fence. I heard of an instance of a fox, hard and long pressed, that took to a rail-fence, and, after walking some distance, made a leap to one side to a hollow stump, in the cavity of which he snugly stowed himself. The ruse succeeded, and the dogs lost the trail; but the hunter, coming up, passed by chance near the stump, when out bounded the fox, his cunning availing him less than he deserved. On another occasion the fox took to the public road, and stepped with great care and precision into a sleigh-track. The hard, polished snow took no imprint of the light foot, and the scent was no doubt less than it would have been on a rougher surface. Maybe, also, the rogue had considered the chances of another sleigh coming along, before the hound, and obliterating the trail entirely.

Audubon tells us of a certain fox, which, when started by the hounds, always managed to elude them at a certain point. Finally the hunter concealed himself in the locality, to discover, if possible, the trick. Presently along came the fox, and, making a leap to one side, ran up the trunk of a fallen tree which had lodged some feet from the ground, and concealed himself in the top. In a few minutes the hounds came up, and in their eagerness passed some distance beyond the point, and then went still farther, looking for the lost trail. Then the fox hastened down, and, taking his back-track, fooled the dogs completely.

I was told of a silver-gray fox in northern New York, which, when pursued by the hounds, would run till it had hunted up another fox, or the fresh trail of one, when it would so manœuvre that the hound would invariably be switched off on the second track.

THE FOX

In cold, dry weather the fox will sometimes elude the hound, at least delay him much, by taking to a bare, plowed field. The hard dry earth seems not to retain a particle of the scent, and the hound gives a loud, long, peculiar bark, to signify he has trouble. It is now his turn to show his wit, which he often does by passing completely around the field, and resuming the trail again where it crosses the fence or a strip of snow.

The fact that any dry, hard surface is unfavorable to the hound suggests, in a measure, the explanation of the wonderful faculty that all dogs in a degree possess to track an animal by the scent of the foot alone. Did you ever think why a dog's nose is always wet? Examine the nose of a foxhound, for instance; how very moist and sensitive! Cause this moisture to dry up, and the dog would be as powerless to track an animal as you are! The nose of the cat, you may observe, is but a little moist, and, as you know, her sense of smell is far inferior to that of the dog. Moisten your own nostrils and lips, and this sense is plainly sharpened. The sweat of a dog's nose, therefore, is no doubt a vital element in its power, and, without taking a very long logical stride, we may infer how much a damp, rough surface aids him in tracking game.

A fox hunt in this country is, of course, quite a different thing from what it is in England, where all the squires and noblemen of a borough, superbly mounted, go riding over the country, guided by the yelling hounds, till the fox is literally run down and murdered. Here the hunter prefers a rough, mountainous country, and, as probably most persons know, takes advantage of the disposition of the fox, when pursued by the hound, to play or circle around a ridge or bold point, and, taking his stand near the run-way, shoots him down.

I recently had the pleasure of a turn with some experienced hunters. As we ascended the ridge toward the mountain, keeping in our ears the uncertain baying of the hounds as they slowly unraveled an old trail, my companions pointed out to me the different run-ways, — a gap in the fence here, a rock just below the brow of the hill there, that tree yonder near the corner of the woods, or the end of that stone wall looking down the side-hill, or commanding a cow-path, or the outlet of a wood-road. A half-wild apple orchard near a cross-road was pointed out as an invariable run-way, where the fox turned toward the mountain again, after having been driven down the ridge. There appeared to be no reason why the foxes should habitually pass any particular point, yet the hunters told me that year after year they took about the same turns, each generation of foxes running through the upper corner of that field, or crossing the valley near yonder stone wall, when pursued by the dog. It seems the fox when he finds himself followed is perpetually tempted to turn in his course, to deflect from a right line, as a person would undoubtedly be under similar circumstances. If he is on this side of the ridge, when he hears the dog break around on his trail he speedily crosses to the other side; if he is in the fields, he takes again to the woods; if in the valley, he hastens to the high land, and evidently enjoys running along the ridge and listening to the dogs, slowly tracing out his course in the fields below. At such times he appears to have but one sense, hearing, and that seems to be reverted toward his pursuers. He is constantly pausing, looking back and listening, and will almost run over the hunter if he stands still, even though not at all concealed.

Animals of this class depend far less upon their sight than upon their hearing and sense of smell. Neither the fox nor the dog is capable of much discrimination with the eye; they seem to see things only in the mass; but with the nose they can analyze and define, and get at the most subtle shades of difference. The fox will not read a man from a stump or a rock, unless he gets his scent, and the dog does not know his master in a crowd until he has smelled him.

On the occasion to which I refer, it was not many minutes after the dogs entered the woods on the side of the mountain before they gave out sharp and eager, and we knew at once that the fox was started. We were then near a point that had been designated as a sure run-way, and hastened to get into position with all speed. For my part I was so taken with the music of the hounds, as it swelled up over the ridge, that I quite forgot the game. I saw one of my companions leveling his gun, and, looking a few rods to the right, saw the fox coming right on to us. I had barely time to note the silly and abashed expression that came over him as he saw us in his path, when he was cut down as by a flash of lightning. The rogue did not appear frightened, but ashamed and out of countenance, as one does when some trick has been played upon him, or when detected in some mischief.

Late in the afternoon, as we were passing through a piece of woods in the valley below, another fox, the third that day, broke from his cover in an old treetop, under our very noses, and drew the fire of three of our party, myself among the number, but, thanks to the interposing trees and limbs, escaped unhurt. Then the dogs took up the trail and there was lively music again. The fox steered through the fields direct for the ridge where we had passed up in the morning. We knew he would take a turn here and then point for the mountain, and two of us, with the hope of cutting him off by the old orchard, through which we were again assured he would surely pass, made a precipitous rush for that point. It was nearly half a mile distant, most of the way up a steep side-hill, and if the fox took the circuit indicated he would probably be there in twelve or fifteen minutes. Running up an angle of 45° seems quite easy work for a four-footed beast like a dog or a fox, but for a two-legged animal like a man it is very heavy and awkward. Before I got halfway up there seemed to be a vacuum all about me, so labored was my breathing, and when I reached the summit my head swam and my knees were about giving out; but pressing on, I had barely time to reach a point in the road abreast of the orchard, when I heard the hounds, and, looking under the trees, saw the fox, leaping high above the weeds and grass, coming straight toward me. He evidently had not got over the first scare, which our haphazard fusillade had given him, and was making unusually quick time. I was armed with a rifle, and said to myself that now was the time to win the laurels I had coveted. For half a day previous I had been practicing on a pumpkin which a patient youth had rolled down a hill for me, and had improved my shot considerably. Now a yellow pumpkin was coming which was not a pumpkin, and for the first time during the day opportunity favored me. I expected the fox to cross the road a few yards below me, but just then I heard him whisk through the grass, and he bounded upon the fence a few yards above. He seemed to cringe as he saw his old enemy, and to depress his fur to half his former dimensions. Three bounds and he had cleared the road, when my bullet tore up the sod beside him, but to this hour I do not know whether I looked at the fox without seeing my gun, or whether I did sight him across its barrel. I only know that I did not distinguish myself in the use of the rifle on that occasion, and went home to wreak my revenge upon another pumpkin; but without much improvement of my skill, for, a few days after, another fox ran under my very nose with perfect impunity. There is something so fascinating in the sudden appearance of the fox that the eye is quite mastered, and, unless the instinct of the sportsman is very strong and quick, the prey will slip through his grasp.

A still hunt rarely brings you in sight of a fox, as his ears are much sharper than yours, and his tread much lighter. But if the fox is mousing in the fields, and you discover him before he does you, you may, the wind favoring, call him within a few paces of you. Secrete yourself behind the fence, or some other object, and squeak as nearly like a mouse as possible. Reynard will hear the sound at an incredible distance. Pricking up his ears, he gets the direction, and comes trotting along as unsuspiciously as can be. I have never had an opportunity to try the experiment, but I know perfectly reliable persons who have. One man, in the pasture getting his cows, called a fox which was too busy mousing to get the first sight, till it jumped upon the wall just over where he sat secreted. Giving a loud whoop and jumping up at the same time, the fox came as near being frightened out of his skin as I suspect a fox ever was.

In trapping for a fox, you get perhaps about as much "fun" and as little fur as in any trapping amusement you can engage in. The one feeling that ever seems present to the mind of Reynard is suspicion. He does not need experience to teach him, but seems to know from the jump that there is such a thing as a trap, and that a trap has a way of grasping a fox's paw that is more frank than friendly. Cornered in a hole or a den, a trap can be set so that the poor creature has the desperate alternative of being caught or starving. He is generally caught, though not till he has braved hunger for a good many days.

But to know all his cunning and shrewdness, bait him in the field, or set your trap by some carcass where he is wont to come. In some cases he will uncover the trap, and leave the marks of his contempt for it in a way you cannot mistake, or else he will not approach within a rod of it. Occasionally, however, he finds in a trapper more than his match, and is fairly caught. When this happens, the trap, which must be of the finest make, is never touched with the bare hand, but, after being thoroughly smoked and greased, is set in a bed of dry ashes or chaff in a remote field, where the fox has been emboldened to dig for several successive nights for morsels of toasted cheese.

A light fall of snow aids the trapper's art and conspires to Reynard's ruin. But how lightly he is caught, when caught at all! barely the end of his toes, or at most a spike through the middle of his foot. I once saw a large painting of a fox struggling with a trap which held him by the hind leg, above the gambrel-joint! A painting alongside of it represented a peasant driving an ox-team from the off-side! A fox would be as likely to be caught above the gambrel-joint as a farmer would to drive his team from the off-side. I knew one that was caught by the tip of the lower jaw. He came nightly, and took the morsel of cheese from the pan of the trap without springing it. A piece was then secured to the pan by a thread, with the result as above stated.

I have never been able to see clearly why the mother fox generally selects a burrow or hole in the open field in which to have her young, except it be, as some hunters maintain, for better security. The young foxes are wont to come out on a warm day, and play like puppies in front of the den. The view being unobstructed on all sides by trees or bushes, in the cover of which danger might approach, they are less liable to surprise and capture. On the slightest sound they disappear in the hole. Those who have watched the gambols of young foxes speak of

them as very amusing, even more arch and playful
than those of kittens, while a spirit profoundly wise
and cunning seems to look out of their young eyes.
The parent fox can never be caught in the den with
them, but is hovering near the woods, which are
always at hand, and by her warning cry or bark
tells them when to be on their guard. She usually
has at least three dens, at no great distance apart,
and moves stealthily in the night with her charge
from one to the other, so as to mislead her enemies.
Many a party of boys, and of men, too, discovering
the whereabouts of a litter, have gone with shovels
and picks, and, after digging away vigorously for
several hours, have found only an empty hole for
their pains. The old fox, finding her secret had been
found out, had waited for darkness, in the cover of
which to transfer her household to new quarters;
or else some old fox-hunter, jealous of the preservation of his game, and getting word of the intended
destruction of the litter, had gone at dusk the
night before, and made some disturbance about
the den, perhaps flashed some powder in its mouth,
— a hint which the shrewd animal knew how to
interpret.

The more scientific aspects of the question may
not be without interest to some of my readers. The
fox belongs to the great order of flesh-eating animals called *Carnivora*, and of the family called
Canidæ, or dogs. The wolf is a kind of wild dog,

and the fox is a kind of wolf. Foxes, unlike wolves, however, never go in packs or companies, but hunt singly. The fox has a kind of bark which suggests the dog, as have all the members of this family. The kinship is further shown by the fact that during certain periods, for the most part in the summer, the dog cannot be made to attack or even to pursue the female fox, but will run from her in the most shamefaced manner, which he will not do in the case of any other animal except a wolf. Many of the ways and manners of the fox, when tamed, are also like the dog's. I once saw a young red fox exposed for sale in the market in Washington. A colored man had him, and said he had caught him out in Virginia. He led him by a small chain, as he would a puppy, and the innocent young rascal would lie on his side and bask and sleep in the sunshine, amid all the noise and chaffering around him, precisely like a dog. He was about the size of a full-grown cat, and there was a bewitching beauty about him that I could hardly resist. On another occasion, I saw a gray fox, about two thirds grown, playing with a dog of about the same size, and by nothing in the manners of either could you tell which was the dog and which the fox.

Some naturalists think there are but two perma nent species of the fox in the United States, namely the gray fox and the red fox, though there are five or six varieties. The gray fox, which is much smaller and less valuable than the red, is the Southern species, and is said to be rarely found north of Maryland, though in certain rocky localities along the Hudson it is common.

In the Southern States this fox is often hunted in the English fashion, namely, on horseback, the riders tearing through the country in pursuit till the animal is run down and caught. This is the only fox that will tree. When too closely pressed, instead of taking to a den or a hole, it climbs beyond the reach of the dogs in some small tree.

The red fox is the Northern species, and is rarely found farther south than the mountainous districts of Virginia. In the Arctic regions it gives place to the Arctic fox, which most of the season is white.

The prairie fox, the cross fox, and the black or silver-gray fox seem only varieties of the red fox, as the black squirrel breeds from the gray, and the black woodchuck is found with the brown. There is little to distinguish them from the red, except the color, though the prairie fox is said to be the larger of the two.

The cross fox is dark brown on its muzzle and extremities, with a cross of red and black on its shoulders and breast, which peculiarity of coloring, and not any trait in its character, gives it its name. It is very rare, and few hunters have ever seen one. The American Fur Company used to obtain annually from fifty to one hundred skins. The

skins formerly sold for twenty-five dollars, though
I believe they now bring only about five dollars.

The black or silver-gray fox is the rarest of all,
and its skin the most valuable. The Indians used
to estimate it equal to forty beaver skins. The great
fur companies seldom collect in a single season
more than four or five skins at any one post. Most
of those of the American Fur Company come from
the head-waters of the Mississippi. One of the
younger Audubons shot one in northern New York.
The fox had been seen and fired at many times by
the hunters of the neighborhood, and had come to
have the reputation of leading a charmed life, and
of being invulnerable to anything but a silver bullet. But Audubon brought her down (for it was a
female) on the second trial. She had a litter of
young in the vicinity, which he also dug out, and
found the nest to hold three black and four red
ones, which fact settled the question with him that
black and red often have the same parentage, and
are in truth the same species.

The color of this fox, in a point-blank view, is
black, but viewed at an angle it is a dark silver-
gray, whence has arisen the notion that the black
and the silver-gray are distinct varieties. The tip
of the tail is always white.

In almost every neighborhood there are traditions
of this fox, and it is the dream of young sportsmen; but I have yet to meet the person who has

seen one. I should go well to the north, into the British Possessions, if I were bent on obtaining a specimen.

One more item from the books. From the fact that in the bone caves in this country skulls of the gray fox are found, but none of the red, it is inferred by some naturalists that the red fox is a descendant from the European species, which it resembles in form but surpasses in beauty, and its appearance on this continent is of comparatively recent date.

TOUCHES OF NATURE

*At the grave of Ralph Waldo Emerson
in Concord, Massachusetts*

EDITOR'S NOTE

"Touches of Nature" appeared in *Galaxy* in September 1876 and was reprinted in *Birds and Poets* in 1877. Burroughs called the essay a "kind of patchwork, or coat of many colors." Divided into seventeen sections, the piece ranges from reflections on the Homeric aspects of technology to the decline of American humor, from Thoreau as a "high-nosed man" to Whitman as a "true naturalist." The diversity of topics was common among earlier writers of natural history before the nature essay became more narrowly focused toward the end of the nineteenth century. Earlier writers like Bartram and Audubon would find Burroughs's reflection on cows and his amusing cameo of a schoolboy as appropriate to a work of natural history as observations about loons and hornets. Equally diverse in this essay are Burroughs's philosophical and scientific concepts. At times he demonstrates an emblematic and teleologic approach toward nature; at other times he accepts nature as amoral and meaningless in human terms. His attitude toward the "balance of nature," an ancient concept found

EDITOR'S NOTE

even in Herodotus and Plato to explain the immutability of species, stands uneasily alongside more Darwinian views that necessarily accept the extinction of species. Most surprising to some readers is Burroughs's dark stance as a modern man unwilling to accept human primacy in the scheme of nature.

F.B.

TOUCHES OF NATURE

I

WHEREVER Nature has commissioned one creature to prey upon another, she has preserved the balance by forewarning that other creature of what she has done. Nature says to the cat, "Catch the mouse," and she equips her for that purpose; but on the selfsame day she says to the mouse, "Be wary, — the cat is watching for you." Nature takes care that none of her creatures have smooth sailing, the whole voyage at least. Why has she not made the mosquito noiseless and its bite itchless? Simply because in that case the odds would be too greatly in its favor. She has taken especial pains to enable the owl to fly softly and silently, because the creatures it preys upon are small and wary, and never venture far from their holes. She has not shown the same caution in the case of the crow, because the crow feeds on dead flesh, or on grubs and beetles, or fruit and grain, that do not need to be approached stealthily. The big fish love to eat up the little fish, and the little fish know it, and, on the very day they are hatched, seek shallow water, and put little sandbars between themselves and their too loving parents.

How easily a bird's tail, or that of any fowl, or in fact any part of the plumage, comes out when the hold of its would-be capturer is upon this alone; and how hard it yields in the dead bird! No doubt there is relaxation in the former case. Nature says to the pursuer, "Hold on," and to the pursued, "Let your tail go." What is the tortuous, zigzag course of those slow-flying moths for but to make it difficult for the birds to snap them up? The skunk is a slow, witless creature, and the fox and lynx love its meat; yet it carries a bloodless weapon that neither likes to face.

I recently heard of an ingenious method a certain other simple and slow-going creature has of baffling its enemy. A friend of mine was walking in the fields when he saw a commotion in the grass a few yards off. Approaching the spot, he found a snake — the common garter snake — trying to swallow a lizard. And how do you suppose the lizard was defeating the benevolent designs of the snake? By simply taking hold of its own tail and making itself into a hoop. The snake went round and round, and could find neither beginning nor end. Who was the old giant that found himself wrestling with Time? This little snake had a tougher customer the other day in the bit of eternity it was trying to swallow.

TOUCHES OF NATURE

The snake itself has not the same wit, because I lately saw a black snake in the woods trying to swallow the garter snake, and he had made some headway, though the little snake was fighting every inch of the ground, hooking his tail about sticks and bushes, and pulling back with all his might, apparently not liking the look of things down there at all. I thought it well to let him have a good taste of his own doctrines, when I put my foot down against further proceedings.

This arming of one creature against another is often cited as an evidence of the wisdom of Nature, but it is rather an evidence of her impartiality. She does not care a fig more for one creature than for another, and is equally on the side of both, or perhaps it would be better to say she does not care a fig for either. Every creature must take its chances, and man is no exception. We can ride if we know how and are going her way, or we can be run over if we fall or make a mistake. Nature does not care whether the hunter slay the beast or the beast the hunter; she will make good compost of them both, and her ends are prospered whichever succeeds.

> "If the red slayer think he slays,
>   Or if the slain think he is slain,
> They know not well the subtle ways
>   I keep, and pass, and turn again."

What is the end of Nature? Where is the end

of a sphere ? The sphere balances at any and every point. So everything in Nature is at the top, and yet no *one* thing is at the top.

She works with reference to no measure of time, no limit of space, and with an abundance of material, not expressed by exhaustless. Did you think Niagara a great exhibition of power? What is that, then, that withdraws noiseless and invisible in the ground about, and of which Niagara is but the lifting of the finger?

Nature is thoroughly selfish, and looks only to her own ends. One thing she is bent upon, and that is keeping up the supply, multiplying endlessly and scattering as she multiplies. Did Nature have in view our delectation when she made the apple, the peach, the plum, the cherry? Undoubtedly; but only as a means to her own private ends. What a bribe or a wage is the pulp of these delicacies to all creatures to come and sow their seed! And Nature has taken care to make the seed indigestible, so that, though the fruit be eaten, the germ is not, but only planted.

God made the crab, but man made the pippin; but the pippin cannot propagate itself, and exists only by violence and usurpation. Bacon says, "It is easier to deceive Nature than to force her," but it seems to me the nurserymen really force her. They cut off the head of a savage and clap on the head of a fine gentleman, and the crab becomes a

Swaar or a Baldwin. Or is it a kind of deception practiced upon Nature, which succeeds only by being carefully concealed? If we could play the same tricks upon her in the human species, how the great geniuses could be preserved and propagated, and the world stocked with them! But what a frightful condition of things that would be! No new men, but a tiresome and endless repetition of the old ones, — a world perpetually stocked with Newtons and Shakespeares!

We say Nature knows best, and has adapted this or that to our wants or to our constitution, — sound to the ear, light and color to the eye; but she has not done any such thing, but has adapted man to these things. The physical cosmos is the mould, and man is the molten metal that is poured into it. The light fashioned the eye, the laws of sound made the ear; in fact, man is the outcome of Nature and not the reverse. Creatures that live forever in the dark have no eyes; and would not any one of our senses perish and be shed, as it were, in a world where it could not be used?

II

It is well to let down our metropolitan pride a little. Man thinks himself at the top, and that the immense display and prodigality of Nature are for him. But they are no more for him than they are for the birds and beasts, and he is no more at the top than they are. He appeared upon the stage when the play had advanced to a certain point, and he will disappear from the stage when the play has reached another point, and the great drama will go on without him. The geological ages, the convulsions and parturition throes of the globe, were to bring him forth no more than the beetles. Is not all this wealth of the seasons, these solar and sidereal influences, this depth and vitality and internal fire, these seas, and rivers, and oceans, and atmospheric currents, as necessary to the life of the ants and worms we tread under foot as to our own? And does the sun shine for me any more than for yon butterfly? What I mean to say is, we cannot put our finger upon this or that and say, Here is the end of Nature. The Infinite cannot be measured. The plan of Nature is so immense, — but she has no plan, no scheme, but to go on and on forever. What is size, what is time, distance, to the Infinite? Nothing. The Infinite knows no time, no space, no great, no small, no beginning, no end.

I sometimes think that the earth and the worlds are a kind of nervous ganglia in an organization of which we can form no conception, or less even than that. If one of the globules of blood that circulate in our veins were magnified enough million times, we might see a globe teeming with life and power. Such is this earth of ours, coursing in the veins of the Infinite. Size is only relative, and

the imagination finds no end to the series either
way.

III

Looking out of the car window one day, I saw
the pretty and unusual sight of an eagle sitting
upon the ice in the river, surrounded by half a
dozen or more crows. The crows appeared as if
looking up to the noble bird and attending his movements. "Are those its young?" asked a gentleman
by my side. How much did that man know — not
about eagles, but about Nature? If he had been
familiar with geese or hens, or with donkeys, he
would not have asked that question. The ancients
had an axiom that he who knew one truth knew all
truths; so much else becomes knowable when one
vital fact is thoroughly known. You have a key,
a standard, and cannot be deceived. Chemistry,
geology, astronomy, natural history, all admit one
to the same measureless interiors.

I heard a great man say that he could see how
much of the theology of the day would fall before
the standard of him who had got even the insects.
And let any one set about studying these creatures
carefully, and he will see the force of the remark.
We learn the tremendous doctrine of metamorphosis
from the insect world; and have not the bee and
the ant taught man wisdom from the first? I was
highly edified the past summer by observing the

ways and doings of a colony of black hornets that
established themselves under one of the projecting
gables of my house. This hornet has the reputation
of being a very ugly customer, but I found it no
trouble to live on the most friendly terms with her.
She was as little disposed to quarrel as I was. She
is indeed the eagle among hornets, and very noble
and dignified in her bearing. She used to come
freely into the house and prey upon the flies. You
would hear that deep, mellow hum, and see the
black falcon poising on wing, or striking here and
there at the flies, that scattered on her approach
like chickens before a hawk. When she had caught
one, she would alight upon some object and proceed to dress and draw her game. The wings were
sheared off, the legs cut away, the bristles trimmed,
then the body thoroughly bruised and broken.
When the work was completed, the fly was rolled
up into a small pellet, and with it under her arm
the hornet flew to her nest, where no doubt in due
time it was properly served up on the royal board.
Every dinner inside these paper walls is a state dinner, for the queen is always present.

I used to mount the ladder to within two or three
feet of the nest and observe the proceedings. I at
first thought the workshop must be inside, — a
place where the pulp was mixed, and perhaps
treated with chemicals; for each hornet, when she
came with her burden of materials, passed into the

nest, and then, after a few moments, emerged again and crawled to the place of building. But I one day stopped up the entrance with some cotton, when no one happened to be on guard, and then observed that, when the loaded hornet could not get inside, she, after some deliberation, proceeded to the unfinished part and went forward with her work. Hence I inferred that maybe the hornet went inside to report and to receive orders, or possibly to surrender her material into fresh hands. Her career when away from the nest is beset with dangers; the colony is never large, and the safe return of every hornet is no doubt a matter of solicitude to the royal mother.

The hornet was the first paper-maker, and holds the original patent. The paper it makes is about like that of the newspaper; nearly as firm, and made of essentially the same material, — woody fibres scraped from old rails and boards. And there is news on it, too, if one could make out the characters.

When I stopped the entrance with cotton, there was no commotion or excitement, as there would have been in the case of yellow-jackets. Those outside went to pulling, and those inside went to pushing and chewing. Only once did one of the outsiders come down and look me suspiciously in the face, and inquire very plainly what my business might be up there. I bowed my head, being at the top of a twenty-foot ladder, and had nothing to say.

The cotton was chewed and moistened about the edges till every fibre was loosened, when the mass dropped. But instantly the entrance was made smaller, and changed so as to make the feat of stopping it more difficult.

IV

There are those who look at Nature from the standpoint of conventional and artificial life,— from parlor windows and through gilt-edged poems, — the sentimentalists. At the other extreme are those who do not look at Nature at all, but are a grown part of her, and look away from her toward the other class, — the backwoodsmen and pioneers, and all rude and simple persons. Then there are those in whom the two are united or merged, — the great poets and artists. In them the sentimentalist is corrected and cured, and the hairy and taciturn frontiersman has had experience to some purpose. The true poet knows more about Nature than the naturalist because he carries her open secrets in his heart. Eckermann could instruct Goethe in ornithology, but could not Goethe instruct Eckermann in the meaning and mystery of the bird? It is my privilege to number among my friends a man who has passed his life in cities amid the throngs of men, who never goes to the woods or to the country, or hunts or fishes, and yet he is the true naturalist. I think he studies the orbs. I think day and night and the stars, and the faces of men and women, have taught him all there is worth knowing.

We run to Nature because we are afraid of man. Our artists paint the landscape because they cannot paint the human face. If we could look into the eyes of a man as coolly as we can into the eyes of an animal, the products of our pens and brushes would be quite different from what they are.

V

But I suspect, after all, it makes but little difference to which school you go, whether to the woods or to the city. A sincere man learns pretty much the same things in both places. The differences are superficial, the resemblances deep and many. The hermit is a hermit, and the poet a poet, whether he grow up in the town or the country. I was forcibly reminded of this fact recently on opening the works of Charles Lamb after I had been reading those of our Henry Thoreau. Lamb cared nothing for nature, Thoreau for little else. One was as attached to the city and the life of the street and tavern as the other to the country and the life of animals and plants. Yet they are close akin. They give out the same tone and are pitched in about the same key. Their methods are the same; so are their quaintness and scorn of rhetoric. Thoreau has the drier humor, as might be expected, and is less stomachic. There is more juice and unction in Lamb, but this he owes to his nationality. Both are essayists who in a less reflective age would have been poets pure and simple. Both were spare, high-nosed men, and I fancy a resemblance even in their portraits. Thoreau is the Lamb of New England fields and woods, and Lamb is the Thoreau of London streets and clubs. There was a willful-ness and perversity about Thoreau, behind which he concealed his shyness and his thin skin, and there was a similar foil in Lamb, though less marked, on account of his good-nature; that was a part of his armor, too.

VI

Speaking of Thoreau's dry humor reminds me how surely the old English unctuous and sympa-thetic humor is dying out or has died out of our literature. Our first notable crop of authors had it, — Paulding, Cooper, Irving, and in a measure Hawthorne, — but our later humorists have it not at all, but in its stead an intellectual quickness and perception of the ludicrous that is not unmixed with scorn.

One of the marks of the great humorist, like Cer-vantes, or Sterne, or Scott, is that he approaches his subject, not through his head merely, but through his heart, his love, his humanity. His

humor is full of compassion, full of the milk of
human kindness, and does not separate him from
his subject, but unites him to it by vital ties. How
Sterne loved Uncle Toby and sympathized with
him, and Cervantes his luckless knight! I fear our
humorists would have made fun of them, would
have shown them up and stood aloof superior, and
"laughed a laugh of merry scorn." Whatever else
the great humorist or poet, or any artist, may be or
do, there is no contempt in his laughter. And this
point cannot be too strongly insisted on in view of
the fact that nearly all our humorous writers seem
impressed with the conviction that their own dignity
and self-respect require them to *look down* upon
what they portray. But it is only little men who
look down upon anything or speak down to anybody.

One sees every day how clear it is that specially
fine, delicate, intellectual persons cannot portray
satisfactorily coarse, common, uncultured characters.
Their attitude is at once scornful and supercilious.
The great man, like Socrates, or Dr. Johnson, or
Abraham Lincoln, is just as surely coarse as he is
fine, but the complaint I make with our humorists
is that they are fine and not coarse in any healthful and manly sense. A great part of the best literature and the best art is of the vital fluids, the
bowels, the chest, the appetites, and is to be read
and judged only through love and compassion. Let
us pray for unction, which is the marrowfat of

humor, and for humility, which is the badge of manhood.

As the voice of the American has retreated from his chest to his throat and nasal passages, so there is danger that his contribution to literature will soon cease to imply any blood or viscera, or healthful carnality, or depth of human and manly affection, and will be the fruit entirely of our toploftical brilliancy and cleverness.

What I complain of is just as true of the essayists and the critics as of the novelists. The prevailing tone here also is born of a feeling of immense superiority. How our lofty young men, for instance, look down upon Carlyle, and administer their masterly rebukes to him! But see how Carlyle treats Burns, or Scott, or Johnson, or Novalis, or any of his heroes. Ay, there's the rub; he makes heroes of them, which is not a trick of small natures. He can say of Johnson that he was "moonstruck," but it is from no lofty height of fancied superiority, but he uses the word as a naturalist uses a term to describe an object he loves.

What we want, and perhaps have got more of than I am ready to admit, is a race of writers who affiliate with their subjects, and enter into them through their blood, their sexuality and manliness, instead of standing apart and criticising them and writing *about* them through mere intellectual cleverness and "smartness."

VII

There is a feeling in heroic poetry, or in a burst of eloquence, that I sometimes catch in quite different fields. I caught it this morning, for instance, when I saw the belated trains go by, and knew how they had been battling with storm, darkness, and distance, and had triumphed. They were due at my place in the night, but did not pass till after eight o'clock in the morning. Two trains coupled together, — the fast mail and the express, — making an immense line of coaches hauled by two engines. They had come from the West, and were all covered with snow and ice, like soldiers with the dust of battle upon them. They had massed their forces, and were now moving with augmented speed, and with a resolution that was epic and grand. Talk about the railroad dispelling the romance from the landscape; if it does, it brings the heroic element in. The moving train is a proud spectacle, especially on stormy and tempestuous nights. When I look out and see its light, steady and unflickering as the planets, and hear the roar of its advancing tread, or its sound diminishing in the distance, I am comforted and made stout of heart. O night, where is thy stay! O space, where is thy victory! Or to see the fast mail pass in the morning is as good as a page of Homer. It quickens one's pulse for all day. It is the Ajax of trains. I hear its defiant, warning whistle, hear it thunder over the bridges, and its sharp, rushing ring among the rocks, and in the winter mornings see its glancing, meteoric lights, or in summer its white form bursting through the silence and the shadows, its plume of smoke lying flat upon its roofs and stretching far behind, — a sight better than a battle. It is something of the same feeling one has in witnessing any wild, free careering in storms, and in floods in nature; or in beholding the charge of an army; or in listening to an eloquent man, or to a hundred instruments of music in full blast, — it is triumph, victory. What is eloquence but mass in motion, — a flood, a cataract, an express train, a cavalry charge? We are literally carried away, swept from our feet, and recover our senses again as best we can.

I experienced the same emotion when I saw them go by with the sunken steamer. The procession moved slowly and solemnly. It was like a funeral cortège, — a long line of grim floats and barges and boxes, with their bowed and solemn derricks, the pall-bearers; and underneath in her watery grave, where she had been for six months, the sunken steamer, partially lifted and borne along. Next day the procession went back again, and the spectacle was still more eloquent. The steamer had been taken to the flats above and raised till her walking-beam was out of water; her bell also was exposed and cleaned and rung, and the wreckers' Herculean

labor seemed nearly over. But that night the winds
and the storms held high carnival. It looked like
preconcerted action on the part of tide, tempest,
and rain to defeat these wreckers, for the elements all pulled together and pulled till cables and
hawser snapped like threads. Back the procession
started, anchors were dragged or lost, immense new
cables were quickly taken ashore and fastened to
trees; but no use: trees were upturned, the cables
stretched till they grew small and sang like harp-
strings, then parted; back, back against the desperate efforts of the men, till within a few feet of
her old grave, when there was a great commotion
among the craft, floats were overturned, enormous
chains parted, colossal timbers were snapped like
pipestems, and, with a sound that filled all the
air, the steamer plunged to the bottom again in
seventy feet of water.

VIII

I am glad to observe that all the poetry of the
midsummer harvesting has not gone out with the
scythe and the whetstone. The line of mowers was
a pretty sight, if one did not sympathize too deeply
with the human backs turned up there to the sun,
and the sound of the whetstone, coming up from the
meadows in the dewy morning, was pleasant music.
But I find the sound of the mowing-machine and
the patent reaper is even more in tune with the

voices of Nature at this season. The characteristic sounds of midsummer are the sharp, whirring crescendo of the cicada or harvest fly, and the rasping, stridulous notes of the nocturnal insects. The mowing-machine repeats and imitates these sounds. 'T is like the hum of a locust or the shuffling of a mighty grasshopper. More than that, the grass and the grain at this season have become hard. The timothy stalk is like a file; the rye straw is glazed with flint; the grasshoppers snap sharply as they fly up in front of you; the bird-songs have ceased; the ground crackles under foot; the eye of day is brassy and merciless; and in harmony with all these things is the rattle of the mower and the hay-tedder.

IX

'T is an evidence of how directly we are related to Nature, that we more or less sympathize with the weather, and take on the color of the day. Goethe said he worked easiest on a high barometer. One is like a chimney that draws well some days and won't draw at all on others, and the secret is mainly in the condition of the atmosphere. Anything positive and decided with the weather is a good omen. A pouring rain may be more auspicious than a sleeping sunshine. When the stove draws well, the fogs and fumes will leave your mind.

I find there is great virtue in the bare ground, and have been much put out at times by those white

angelic days we have in winter, such as Whittier
has so well described in these lines: —

> "Around the glistening wonder bent
>   The blue walls of the firmament;
> No cloud above, no earth below,
>   A universe of sky and snow."

On such days my spirit gets snow-blind; all
things take on the same color, or no color; my
thought loses its perspective; the inner world is a
blank like the outer, and all my great ideals are
wrapped in the same monotonous and expression-
less commonplace. The blackest of black days are
better.

Why does snow so kill the landscape and blot out
our interest in it? Not merely because it is cold,
and the symbol of death, — for I imagine as many
inches of apple blossoms would have about the same
effect, — but because it expresses nothing. White is
a negative; a perfect blank. The eye was made for
color, and for the earthy tints, and, when these are
denied it, the mind is very apt to sympathize and
to suffer also.

Then when the sap begins to mount in the trees,
and the spring languor comes, does not one grow
restless indoors? The sun puts out the fire, the
people say, and the spring sun certainly makes one's
intellectual light grow dim. Why should not a man
sympathize with the seasons and the moods and

phases of Nature? He is an apple upon this tree, or rather he is a babe at this breast, and what his great mother feels affects him also.

X

I have frequently been surprised, in late fall and early winter, to see how unequal or irregular was the encroachment of the frost upon the earth. If there is suddenly a great fall in the mercury, the frost lays siege to the soil and effects a lodgment here and there, and extends its conquests gradually. At one place in the field you can easily run your staff through into the soft ground, when a few rods farther on it will be as hard as a rock. A little covering of dry grass or leaves is a great protection. The moist places hold out long, and the spring runs never freeze. You find the frost has gone several inches into the plowed ground, but on going to the woods, and poking away the leaves and débris under the hemlocks and cedars, you find there is no frost at all. The Earth freezes her ears and toes and naked places first, and her body last.

If heat were visible, or if we should represent it say by smoke, then the December landscape would present a curious spectacle. We should see the smoke lying low over the meadows, thickest in the hollows and moist places, and where the turf is oldest and densest. It would cling to the fences and ravines. Under every evergreen tree we should

see the vapor rising and filling the branches, while
the woods of pine and hemlock would be blue with
it long after it had disappeared from the open country. It would rise from the tops of the trees, and be
carried this way and that with the wind. The valleys
of the great rivers, like the Hudson, would overflow
with it. Large bodies of water become regular magazines in which heat is stored during the summer,
and they give it out again during the fall and early
winter. The early frosts keep well back from the
Hudson, skulking behind the ridges, and hardly
come over in sight at any point. But they grow
bold as the season advances, till the river's fires, too,
are put out and Winter covers it with his snows.

XI

One of the strong and original strokes of Nature
was when she made the loon. It is always refreshing to contemplate a creature so positive and characteristic. He is the great diver and flyer under
water. The loon is the *genius loci* of the wild northern lakes, as solitary as they are. Some birds represent the majesty of nature, like the eagles; others
its ferocity, like the hawks; others its cunning,
like the crow; others its sweetness and melody,
like the song-birds. The loon represents its wildness and solitariness. It is cousin to the beaver. It
has the feathers of a bird and the fur of an animal,
and the heart of both. It is as quick and cunning

as it is bold and resolute. It dives with such marvelous quickness that the shot of the gunner get there just in time "to cut across a circle of descending tail feathers and a couple of little jets of water flung upward by the web feet of the loon." When disabled so that it can neither dive nor fly, it is said to face its foe, look him in the face with its clear, piercing eye, and fight resolutely till death. The gunners say there is something in its wailing, piteous cry, when dying, almost human in its agony. The loon is, in the strictest sense, an aquatic fowl. It can barely walk upon the land, and one species at least cannot take flight from the shore. But in the water its feet are more than feet, and its wings more than wings. It plunges into this denser air and flies with incredible speed. Its head and beak form a sharp point to its tapering neck. Its wings are far in front and its legs equally far in the rear, and its course through the crystal depths is like the speed of an arrow. In the northern lakes it has been taken forty feet under water upon hooks baited for the great lake trout. I had never seen one till last fall, when one appeared on the river in front of my house. I knew instantly it was the loon. Who could not tell a loon a half mile or more away, though he had never seen one before? The river was like glass, and every movement of the bird as it sported about broke the surface into ripples, that revealed it far and wide. Presently a boat shot out from shore, and went ripping up the surface toward the loon. The creature at once seemed to divine the intentions of the boatman, and sidled off obliquely, keeping a sharp lookout as if to make sure it was pursued. A steamer came down and passed between them, and when the way was again clear, the loon was still swimming on the surface. Presently it disappeared under the water, and the boatman pulled sharp and hard. In a few moments the bird reappeared some rods farther on, as if to make an observation. Seeing it was being pursued, and no mistake, it dived quickly, and, when it came up again, had gone many times as far as the boat had in the same space of time. Then it dived again, and distanced its pursuer so easily that he gave over the chase and rested upon his oars. But the bird made a final plunge, and, when it emerged upon the surface again, it was over a mile away. Its course must have been, and doubtless was, an actual flight under water, and half as fast as the crow flies in the air.

The loon would have delighted the old poets. Its wild, demoniac laughter awakens the echoes on the solitary lakes, and its ferity and hardiness are kindred to those robust spirits.

XII

One notable difference between man and the four-footed animals which has often occurred to me is in the eye, and the greater perfection, or rather supremacy, of the sense of sight in the human species. All the animals — the dog, the fox, the wolf, the deer, the cow, the horse — depend mainly upon the senses of hearing and smell. Almost their entire powers of discrimination are confined to these two senses. The dog picks his master out of the crowd by smell, and the cow her calf out of the herd. Sight is only partial recognition. The question can only be settled beyond all doubt by the aid of the nose. The fox, alert and cunning as he is, will pass within a few yards of the hunter and not know him from a stump. A squirrel will run across your lap, and a marmot between your feet, if you are motionless. When a herd of cattle see a strange object, they are not satisfied till each one has sniffed it; and the horse is cured of his fright at the robe, or the meal-bag, or other object, as soon as he can be induced to smell it. There is a great deal of speculation in the eye of an animal, but very little science. Then you cannot catch an animal's eye; he looks at you, but not *into* your eye. The dog directs his gaze toward your face, but, for aught you can tell, it centres upon your mouth or nose. The same with your horse or cow. Their eye is vague and indefinite.

Not so with the birds. The bird has the human eye in its clearness, its power, and its supremacy over the other senses. How acute their sense of smell may be is uncertain; their hearing is sharp enough, but their vision is the most remarkable. A crow or a hawk, or any of the larger birds, will not mistake you for a stump or a rock, stand you never so still amid the bushes. But they cannot separate you from your horse or team. A hawk reads a man on horseback as one animal, and reads it as a horse. None of the sharp-scented animals could be thus deceived.

The bird has man's brain also in its size. The brain of a song-bird is even much larger in proportion than that of the greatest human monarch, and its life is correspondingly intense and high-strung. But the bird's eye is superficial. It is on the outside of his head. It is round, that it may take in a full circle at a glance.

All the quadrupeds emphasize their direct forward gaze by a corresponding movement of the ears, as if to supplement and aid one sense with another. But man's eye seldom needs the confirmation of his ear, while it is so set, and his head so poised, that his look is forcible and pointed without being thus seconded.

XIII

I once saw a cow that had lost her cud. How forlorn and desolate and sick at heart that cow looked! No more rumination, no more of that second and finer mastication, no more of that sweet and juicy reverie under the spreading trees, or in the stall. Then the farmer took an elder and scraped the bark and put something with it, and made the cow a cud, and, after due waiting, the experiment took, a response came back, and the mysterious machinery was once more in motion, and the cow was herself again.

Have you, O poet, or essayist, or story-writer, never lost your cud, and wandered about days and weeks without being able to start a single thought or an image that tasted good, — your literary appetite dull or all gone, and the conviction daily growing that it was all over with you in that direction? A little elder-bark, something fresh and bitter from the woods, is about the best thing you can take.

XIV

Notwithstanding what I have elsewhere said about the desolation of snow, when one looks closely it is little more than a thin veil after all, and takes and repeats the form of whatever it covers. Every path through the fields is just as plain as before. On every hand the ground sends tokens, and the curves and slopes are not of the snow, but of the earth beneath. In like manner the rankest vegetation hides the ground less than we think. Looking across a wide valley in the month of July, I have noted that the fields, except the meadows, had a ruddy tinge, and that corn, which near at hand seemed to completely envelop the soil, at that distance gave only a slight shade of green. The color of the ground everywhere predominated, and I doubt not that, if we could see the earth from a point sufficiently removed, as from the moon, its ruddy hue, like that of Mars, would alone be visible.

What is a man but a miniature earth, with many disguises in the way of manners, possessions, dissemblances? Yet through all — through all the work of his hands and all the thoughts of his mind — how surely the ground quality of him, the fundamental hue, whether it be this or that, makes itself felt and is alone important!

XV

Men follow their noses, it is said. I have wondered why the Greek did not follow his nose in architecture, — did not copy those arches that spring from it as from a pier, and support his brow, — but always and everywhere used the post and the lintel. There was something in that face that has never reappeared in the human countenance. I am thinking especially of that straight, strong profile. Is it really godlike, or is this impression the result of association? But any suggestion or reminiscence of it in the modern face at once gives one the idea of strength. It is a face strong in the loins, or it suggests a high, elastic instep. It is the face of order and proportion. Those arches are the symbols of law and self-control. The point of greatest

interest is the union of the nose with the brow, —
that strong, high embankment; it makes the bridge
from the ideal to the real sure and easy. All the
Greek's ideas passed readily into form. In the modern face the arches are more or less crushed, and
the nose is severed from the brow, — hence the abstract and the analytic ; hence the preponderance
of the speculative intellect over creative power.

XVI

I have thought that the boy is the only true lover
of Nature, and that we, who make such a dead set
at studying and admiring her, come very wide of the
mark. "The nonchalance of a boy who is sure of
his dinner," says our Emerson, "is the healthy attitude of humanity." The boy is a part of Nature;
he is as indifferent, as careless, as vagrant as she.
He browses, he digs, he hunts, he climbs, he hal-
loes, he feeds on roots and greens and mast. He
uses things roughly and without sentiment. The
coolness with which boys will drown dogs or cats,
or hang them to trees, or murder young birds, or
torture frogs or squirrels, is like Nature's own mer-
cilessness.

Certain it is that we often get some of the best
touches of nature from children. Childhood is a
world by itself, and we listen to children when they
frankly speak out of it with a strange interest.
There is such a freedom from responsibility and

from worldly wisdom, — it is heavenly wisdom. There is no sentiment in children, because there is no ruin; nothing has gone to decay about them yet, — not a leaf or a twig. Until he is well into his teens, and sometimes later, a boy is like a bean-pod before the fruit has developed, — indefinite, succulent, rich in possibilities which are only vaguely outlined. He is a pericarp merely. How rudimental are all his ideas! I knew a boy who began his school composition on swallows by saying there were two kinds of swallows, — chimney swallows and swallows.

Girls come to themselves sooner; are indeed, from the first, more definite and "translatable."

XVII

Who will write the natural history of the boy? One of the first points to be taken account of is his clannishness. The boys of one neighborhood are always pitted against those of an adjoining neighborhood, or of one end of the town against those of the other end. A bridge, a river, a railroad track, are always boundaries of hostile or semi-hostile tribes. The boys that go up the road from the country school hoot derisively at those that go down the road, and not infrequently add the insult of stones ; and the down-roaders return the hooting and the missiles with interest.

Often there is open war, and the boys meet and

have regular battles. A few years since, the boys of two rival towns on opposite sides of the Ohio River became so belligerent that the authorities had to interfere. Whenever an Ohio boy was caught on the West Virginia side of the river, he was unmercifully beaten; and when a West Virginia boy was discovered on the Ohio side, he was pounced upon in the same manner. One day a vast number of boys, about one hundred and fifty on a side, met by appointment upon the ice and engaged in a pitched battle. Every conceivable missile was used, including pistols. The battle, says the local paper, raged with fury for about two hours. One boy received a wound behind the ear, from the effects of which he died the next morning. More recently the boys of a large manufacturing town of New Jersey were divided into two hostile clans that came into frequent collision. One Saturday both sides mustered their forces, and a regular fight ensued, one boy here also losing his life from the encounter.

Every village and settlement is at times the scene of these youthful collisions. When a new boy appears in the village, or at the country school, how the other boys crowd around him and take his measure, or pick at him and insult him to try his mettle!

I knew a boy, twelve or thirteen years old, who was sent to help a drover with some cattle as far as a certain village ten miles from his home. After the place was reached, and while the boy was eating his cracker and candies, he strolled about the village, and fell in with some other boys playing upon a bridge. In a short time a large number of children of all sizes had collected upon the bridge. The new-comer was presently challenged by the boys of his own age to jump with them. This he readily did, and cleared their farthest mark. Then he gave them a sample of his stone-throwing, and at this pastime he also far surpassed his competitors. Before long, the feeling of the crowd began to set against him, showing itself first in the smaller fry, who began half playfully to throw pebbles and lumps of dry earth at him. Then they would run up slyly and strike him with sticks. Presently the large ones began to tease him in like manner, till the contagion of hostility spread, and the whole pack was arrayed against the strange boy. He kept them at bay for a few moments with his stick, till, the feeling mounting higher and higher, he broke through their ranks, and fled precipitately toward home, with the throng of little and big at his heels. Gradually the girls and smaller boys dropped behind, till at the end of the first fifty rods only two boys of about his own size, with wrath and determination in their faces, kept up the pursuit. But to these he added the final insult of beating them at running also, and reached, much blown, a point beyond which they refused to follow.

The world the boy lives in is separate and distinct from the world the man lives in. It is a world inhabited only by boys. No events are important or of any moment save those affecting boys. How they ignore the presence of their elders on the street, shouting out their invitations, their appointments, their pass-words from our midst, as from the veriest solitude! They have peculiar calls, whistles, signals, by which they communicate with each other at long distances, like birds or wild creatures. And there is as genuine a wildness about these notes and calls as about those of a fox or a coon.

The boy is a savage, a barbarian, in his taste, — devouring roots, leaves, bark, unripe fruit ; and in the kind of music or discord he delights in, — of harmony he has no perception. He has his fashions that spread from city to city. In one of our large cities the rage at one time was an old tin can with a string attached, out of which they tortured the most savage and ear-splitting discords. The police were obliged to interfere and suppress the nuisance. On another occasion, at Christmas, they all came forth with tin horns, and nearly drove the town distracted with the hideous uproar.

Another savage trait of the boy is his untruthfulness. Corner him, and the chances are ten to one he will lie his way out. Conscience is a plant of slow growth in the boy. If caught in one lie, he invents another. I know a boy who was in the habit of eating apples in school. His teacher finally caught him in the act, and, without removing his eye from him, called him to the middle of the floor.

"I saw you this time," said the teacher.

"Saw me what?" said the boy innocently.

"Bite that apple," replied the teacher.

"No, sir," said the rascal.

"Open your mouth;" and from its depths the teacher, with his thumb and finger, took out the piece of apple.

"Did n't know it was there," said the boy, unabashed.

Nearly all the moral sentiment and graces are late in maturing in the boy. He has no proper self-respect till past his majority. Of course there are exceptions, but they are mostly windfalls. The good boys die young. We lament the wickedness and thoughtlessness of the young vagabonds at the same time that we know it is mainly the acridity and bitterness of the unripe fruit that we are lamenting.

APRIL

Vassar students at Slabsides in 1904

EDITOR'S NOTE

"April" is Burroughs's hymn to his favorite month of the year. "I am always sad to see April go," he once wrote. "It is the mother month to me." Excitement enters his journal each year as he records the signs of the approaching month. As a naturalist noted primarily for his sense of sight, Burroughs reveals the acuteness of his other senses as he responds to the sounds and smells of April. In the edition of this essay prepared for the volume *Birds and Poets,* published in 1877 and reprinted here, Burroughs added a section on the sound he identified most intensely with the month, not that of spring birds but of frogs piping in the marshes and signaling the passing of winter. As a transitional month, April is a season of change and movement so attractive to Burroughs's aesthetic sense. Containing the polarities of winter and summer, April is both bitter and sweet, a month of uncloying energy. "There is no month oftener on the tongues of the poets than April," Burroughs wrote in the original version of the essay published, appropriately, in the April issue of *Scribner's* in 1877.

EDITOR'S NOTE

Burroughs expanded this discussion of April poetry into a separate essay also printed in *Birds and Poets*. Burroughs found it fit that Emerson died in April, the month of Shakespeare's birth and death. As it happened, Burroughs himself was born and buried on his family farm on the third day of April.

F.B.

APRIL

IF we represent the winter of our northern climate by a rugged snow-clad mountain, and summer by a broad fertile plain, then the intermediate belt, the hilly and breezy uplands, will stand for spring, with March reaching well up into the region of the snows, and April lapping well down upon the greening fields and unloosened currents, not beyond the limits of winter's sallying storms, but well within the vernal zone, — within the reach of the warm breath and subtle, quickening influences of the plain below. At its best, April is the tenderest of tender salads made crisp by ice or snow water. Its type is the first spear of grass. The senses — sight, hearing, smell — are as hungry for its delicate and almost spiritual tokens as the cattle are for the first bite of its fields. How it touches one and makes him both glad and sad! The voices of the arriving birds, the migrating fowls, the clouds of pigeons sweeping across the sky or filling the woods, the elfin horn of the first honey-bee venturing abroad in the middle of the day, the clear piping of the little frogs in the marshes at sundown, the camp-fire in the sugar-bush, the smoke seen afar rising over the trees, the tinge of green that comes so suddenly on the sunny knolls and slopes, the full translucent streams, the waxing and warming sun, — how these things and others like them are noted by the eager eye and ear! April is my natal month, and I am born again into new delight and new surprises at each return of it. Its name has an indescribable charm to me. Its two syllables are like the calls of the first birds, — like that of the phœbe-bird, or of the meadowlark. Its very snows are fertilizing, and are called the poor man's manure.

Then its odors! I am thrilled by its fresh and indescribable odors, — the perfume of the bursting sod, of the quickened roots and rootlets, of the mould under the leaves, of the fresh furrows. No other month has odors like it. The west wind the other day came fraught with a perfume that was to the sense of smell what a wild and delicate strain of music is to the ear. It was almost transcendental. I walked across the hill with my nose in the air taking it in. It lasted for two days. I imagined it came from the willows of a distant swamp, whose catkins were affording the bees their first pollen; or did it come from much farther, — from beyond the horizon, the accumulated breath of innumerable farms and budding forests? The main characteristic of these April odors is their uncloying freshness. They are not sweet, they are oftener bitter, they are penetrating and lyrical. I know well the odors of May and June, of the world of meadows and orchards bursting into bloom, but they are not so ineffable and immaterial and so stimulating to the sense as the incense of April.

The season of which I speak does not correspond with the April of the almanac in all sections of our vast geography. It answers to March in Virginia and Maryland, while in parts of New York and New England it laps well over into May. It begins when the partridge drums, when the hyla pipes, when the shad start up the rivers, when the grass greens in the spring runs, and it ends when the leaves are unfolding and the last snowflake dissolves in midair. It may be the first of May before the first swallow appears, before the whip-poor-will is heard, before the wood thrush sings; but it is April as long as there is snow upon the mountains, no matter what the almanac may say. Our April is, in fact, a kind of Alpine summer, full of such contrasts and touches of wild, delicate beauty as no other season affords. The deluded citizen fancies there is nothing enjoyable in the country till June, and so misses the freshest, tenderest part. It is as if one should miss strawberries and begin his fruit-eating with melons and peaches. These last are good, — supremely so, they are melting and luscious, — but nothing so thrills and penetrates the taste, and wakes up and teases the papillæ of the tongue, as the uncloying strawberry. What midsummer sweetness half so distracting as its brisk sub-acid flavor, and what splendor of full-leaved June can stir the blood like the best of leafless April?

One characteristic April feature, and one that delights me very much, is the perfect emerald of the spring runs while the fields are yet brown and sere, — strips and patches of the most vivid velvet green on the slopes and in the valleys. How the eye grazes there, and is filled and refreshed! I had forgotten what a marked feature this was until I recently rode in an open wagon for three days through a mountainous, pastoral country, remarkable for its fine springs. Those delicious green patches are yet in my eye. The fountains flowed with May. Where no springs occurred, there were hints and suggestions of springs about the fields and by the roadside in the freshened grass, — sometimes overflowing a space in the form of an actual fountain. The water did not quite get to the surface in such places, but sent its influence.

The fields of wheat and rye, too, how they stand out of the April landscape, — great green squares on a field of brown or gray!

Among April sounds there is none more welcome or suggestive to me than the voice of the little frogs piping in the marshes. No bird-note can surpass it as a spring token; and as it is not mentioned, to my knowledge, by the poets and writers of other lands,

APRIL

I am ready to believe it is characteristic of our season alone. You may be sure April has really come when this little amphibian creeps out of the mud and inflates its throat. We talk of the bird inflating its throat, but you should see this tiny minstrel inflate *its* throat, which becomes like a large bubble, and suggests a drummer-boy with his drum slung very high. In this drum, or by the aid of it, the sound is produced. Generally the note is very feeble at first, as if the frost was not yet all out of the creature's throat, and only one voice will be heard, some prophet bolder than all the rest, or upon whom the quickening ray of spring has first fallen. And it often happens that he is stoned for his pains by the yet unpacified element, and is compelled literally to " shut up " beneath a fall of snow or a heavy frost. Soon, however, he lifts up his voice again with more confidence, and is joined by others and still others, till in due time, say toward the last of the month, there is a shrill musical uproar, as the sun is setting, in every marsh and bog in the land. It is a plaintive sound, and I have heard people from the city speak of it as lonesome and depressing, but to the lover of the country it is a pure spring melody. The little piper will sometimes climb a bulrush, to which he clings like a sailor to a mast, and send forth his shrill call. There is a Southern species, heard when you have reached the Potomac, whose note is far more harsh and crackling. To stand on the verge of a swamp vocal with these, pains and stuns the ear. The call of the Northern species is far more tender and musical.[1]

Then is there anything like a perfect April morning? One hardly knows what the sentiment of it is, but it is something very delicious. It is youth and hope. It is a new earth and a new sky. How the air transmits sounds, and what an awakening, prophetic character all sounds have! The distant barking of a dog, or the lowing of a cow, or the crowing of a cock, seems from out the heart of Nature, and to be a call to come forth. The great sun appears to have been reburnished, and there is something in his first glance above the eastern hills, and the way his eye-beams dart right and left and smite the rugged mountains into gold, that quickens the pulse and inspires the heart.

Across the fields in the early morning I hear some of the rare April birds, — the chewink and the brown thrasher. The robin, the bluebird, the song sparrow, the phœbe-bird, come in March; but these two ground-birds are seldom heard till toward the last of April. The ground-birds are all tree-singers or air-singers; they must have an elevated stage to speak from. Our long-tailed thrush, or thrasher, like its congeners the catbird and the mockingbird, delights in a high branch of some solitary tree,

[1] The Southern species is called the green hyla. I have since heard them in my neighborhood on the Hudson.

whence it will pour out its rich and intricate warble
for an hour together. This bird is the great American chipper. There is no other bird that I know
of that can chip with such emphasis and military
decision as this yellow-eyed songster. It is like the
click of a giant gunlock. Why is the thrasher so
stealthy? It always seems to be going about on tip-
toe. I never knew it to steal anything, and yet it
skulks and hides like a fugitive from justice. One
never sees it flying aloft in the air and traversing the
world openly, like most birds, but it darts along
fences and through bushes as if pursued by a guilty
conscience. Only when the musical fit is upon it
does it come up into full view, and invite the world
to hear and behold.

The chewink is a shy bird also, but not stealthy.
It is very inquisitive, and sets up a great scratching
among the leaves, apparently to attract your attention. The male is perhaps the most conspicuously
marked of all the ground-birds except the bobolink,
being black above, bay on the sides, and white beneath. The bay is in compliment to the leaves he
is forever scratching among, — they have rustled
against his breast and sides so long that these parts
have taken their color; but whence come the white
and the black? The bird seems to be aware that his
color betrays him, for there are few birds in the
woods so careful about keeping themselves screened
from view. When in song, its favorite perch is the

top of some high bush near to cover. On being disturbed at such times, it pitches down into the brush and is instantly lost to view.

This is the bird that Thomas Jefferson wrote to Wilson about, greatly exciting the latter's curiosity. Wilson was just then upon the threshold of his career as an ornithologist, and had made a drawing of the Canada jay which he sent to the President. It was a new bird, and in reply Jefferson called his attention to a "curious bird" which was everywhere to be heard, but scarcely ever to be seen. He had for twenty years interested the young sportsmen of his neighborhood to shoot one for him, but without success. "It is in all the forests, from spring to fall," he says in his letter, "and never but on the tops of the tallest trees, from which it perpetually serenades us with some of the sweetest notes, and as clear as those of the nightingale. I have followed it for miles, without ever but once getting a good view of it. It is of the size and make of the mockingbird, lightly thrush-colored on the back, and a grayish white on the breast and belly. Mr. Randolph, my son-in-law, was in possession of one which had been shot by a neighbor," etc. Randolph pronounced it a flycatcher, which was a good way wide of the mark. Jefferson must have seen only the female, after all his tramp, from his description of the color; but he was doubtless following his own great thoughts more than the bird, else he would have had an earlier view. The bird was not a new one, but was well known then as the ground-robin. The President put Wilson on the wrong scent by his erroneous description, and it was a long time before the latter got at the truth of the case. But Jefferson's letter is a good sample of those which specialists often receive from intelligent persons who have seen or heard something in their line very curious or entirely new, and who set the man of science agog by a description of the supposed novelty, — a description that generally fits the facts of the case about as well as your coat fits the chair-back. Strange and curious things in the air, and in the water, and in the earth beneath, are seen every day except by those who are looking for them, namely, the naturalists. When Wilson or Audubon gets his eye on the unknown bird, the illusion vanishes, and your phenomenon turns out to be one of the commonplaces of the fields or woods.

A prominent April bird, that one does not have to go to the woods or away from his own door to see and hear, is the hardy and ever-welcome meadow-lark. What a twang there is about this bird, and what vigor! It smacks of the soil. It is the winged embodiment of the spirit of our spring meadows. What emphasis in its "z-d-t, z-d-t," and what character in its long, piercing note! Its straight, taper-ing, sharp beak is typical of its voice. Its note goes like a shaft from a crossbow; it is a little too sharp and piercing when near at hand, but, heard in the proper perspective, it is eminently melodious and pleasing. It is one of the major notes of the fields at this season. In fact, it easily dominates all others. "*Spring o' the year! spring o' the year!*" it says, with a long-drawn breath, a little plaintive, but not complaining or melancholy. At times it indulges in something much more intricate and lark-like while hovering on the wing in midair, but a song is beyond the compass of its instrument, and the attempt usually ends in a breakdown. A clear, sweet, strong, high-keyed note, uttered from some knoll or rock, or stake in the fence, is its proper vocal performance. It has the build and walk and flight of the quail and the grouse. It gets up before you in much the same manner, and falls an easy prey to the crack shot. Its yellow breast, surmounted by a black crescent, it need not be ashamed to turn to the morning sun, while its coat of mottled gray is in perfect keeping with the stubble amid which it walks.

The two lateral white quills in its tail seem strictly in character. These quills spring from a dash of scorn and defiance in the bird's make-up. By the aid of these, it can almost emit a flash as it struts about the fields and jerks out its sharp notes. They give a rayed, a definite and piquant expression to its movements. This bird is not properly a lark, but a starling, say the ornithologists, though it is lark-like in its habits, being a walker and entirely a

ground-bird. Its color also allies it to the true lark.
I believe there is no bird in the English or European
fields that answers to this hardy pedestrian of our
meadows. He is a true American, and his note one
of our characteristic April sounds.

Another marked April note, proceeding sometimes from the meadows, but more frequently from
the rough pastures and borders of the woods, is the
call of the high-hole, or golden-shafted woodpecker.
It is quite as strong as that of the meadowlark, but
not so long-drawn and piercing. It is a succession of short notes rapidly uttered, as if the bird
said " *if-if-if-if-if-if-if*." The notes of the ordinary
downy and hairy woodpeckers suggest, in some way,
the sound of a steel punch; but that of the high-
hole is much softer, and strikes on the ear with real
springtime melody. The high-hole is not so much
a wood-pecker as he is a ground-pecker. He subsists
largely on ants and crickets, and does not appear till
they are to be found.

In Solomon's description of spring, the voice of
the turtle is prominent, but our turtle, or mourning
dove, though it arrives in April, can hardly be said
to contribute noticeably to the open-air sounds. Its
call is so vague, and soft, and mournful, — in fact,
so remote and diffused, — that few persons ever hear
it at all.

Such songsters as the cow blackbird are noticeable
at this season, though they take a back seat a little

later. It utters a peculiarly liquid April sound. Indeed, one would think its crop was full of water, its notes so bubble up and regurgitate, and are delivered with such an apparent stomachic contraction. This bird is the only feathered polygamist we have. The females are greatly in excess of the males, and the latter are usually attended by three or four of the former. As soon as the other birds begin to build, they are on the *qui vive*, prowling about like gypsies, not to steal the young of others, but to steal their eggs into other birds' nests, and so shirk the labor and responsibility of hatching and rearing their own young. As these birds do not mate, and as therefore there can be little or no rivalry or competition between the males, one wonders — in view of Darwin's teaching — why one sex should have brighter and richer plumage than the other, which is the fact. The males are easily distinguished from the dull and faded females by their deep glossy-black coats.

The April of English literature corresponds nearly to our May. In Great Britain, the swallow and the cuckoo usually arrive by the middle of April; with us, their appearance is a week or two later. Our April, at its best, is a bright, laughing face under a hood of snow, like the English March, but presenting sharper contrasts, a greater mixture of smiles and tears and icy looks than are known to our ancestral climate. Indeed, Winter sometimes retraces his steps in this month, and unburdens himself of the snows that the previous cold has kept back; but we are always sure of a number of radiant, equable days, — days that go before the bud, when the sun embraces the earth with fervor and determination. How his beams pour into the woods till the mould under the leaves is warm and emits an odor! The waters glint and sparkle, the birds gather in groups, and even those unused to singing find a voice. On the streets of the cities, what a flutter, what bright looks and gay colors! I recall one preëminent day of this kind last April. I made a note of it in my note-book. The earth seemed suddenly to emerge from a wilderness of clouds and chilliness into one of these blue sunlit spaces. How the voyagers rejoiced ! Invalids came forth, old men sauntered down the street, stocks went up, and the political outlook brightened.

Such days bring out the last of the hibernating animals. The woodchuck unrolls and creeps out of his den to see if his clover has started yet. The torpidity leaves the snakes and the turtles, and they come forth and bask in the sun. There is nothing so small, nothing so great, that it does not respond to these celestial spring days, and give the pendulum of life a fresh start.

April is also the month of the new furrow. As soon as the frost is gone and the ground settled, the plow is started upon the hill, and at each bout I see its brightened mould-board flash in the sun. Where the last remnants of the snowdrift lingered yesterday the plow breaks the sod to-day. Where the drift was deepest the grass is pressed flat, and there is a deposit of sand and earth blown from the fields to windward. Line upon line the turf is reversed, until there stands out of the neutral landscape a ruddy square visible for miles, or until the breasts of the broad hills glow like the breasts of the robins.

Then who would not have a garden in April? to rake together the rubbish and burn it up, to turn over the renewed soil, to scatter the rich compost, to plant the first seed, or bury the first tuber! It is not the seed that is planted, any more than it is I that is planted; it is not the dry stalks and weeds that are burned up, any more than it is my gloom and regrets that are consumed. An April smoke makes a clean harvest.

I think April is the best month to be born in. One is just in time, so to speak, to catch the first train, which is made up in this month. My April chickens always turn out best. They get an early start; they have rugged constitutions. Late chickens cannot stand the heavy dews, or withstand the predaceous hawks. In April all nature starts with you. You have not come out of your hibernaculum too early or too late; the time is ripe, and, if you do not keep pace with the rest, why, the fault is not in the season.

SPECKLED TROUT

Trout fishing in Neversink Creek

EDITOR'S NOTE

"I have been a seeker of trout from my boyhood," Burroughs writes in "Speckled Trout," one of the numerous essays in which over the course of his career he records his lifelong delight in trout fishing. The essay recounts his first fishing trip to the Neversink in 1869 and was published in the *Atlantic* in October 1870. When Burroughs prepared the essay for publication in book form nine years later, he left out some details about the actual experience, such as his shooting a robin to eat, his difficulties in sleeping, his headaches on the first day, and his adventure of catching seventy-five trout from a dugout during a thunderstorm. "It was such fun! Sometimes I would haul in two at a time, as I had two flies on my line." To the many mundane and even unappealing incidents he retained in the essay he added a long introductory section on trout fishing as a poetic venture and a way of life. As the opening sentence might suggest, the essay is one of four chapters about hiking and fishing that appear in *Locusts and Wild Honey* (1879), a book whose title was suggested by Burroughs's good friend and

EDITOR'S NOTE

companion Aaron Johns, who himself appears in all four camping and fishing sketches. In the preface to this book of youthful adventure, Burroughs hopes that the title "carries with it a suggestion of the wild and delectable in nature, of the free and ungarnered harvests which the wilderness everywhere affords to the observing eye and ear."

F.B.

SPECKLED TROUT

I

THE legend of the wary trout, hinted at in the last sketch, is to be further illustrated in this and some following chapters. We shall get at more of the meaning of those dark water-lines, and I hope, also, not entirely miss the significance of the gold and silver spots and the glancing iridescent hues. The trout is dark and obscure above, but behind this foil there are wondrous tints that reward the believing eye. Those who seek him in his wild remote haunts are quite sure to get the full force of the sombre and uninviting aspects, — the wet, the cold, the toil, the broken rest, and the huge, savage, uncompromising nature, — but the true angler sees farther than these, and is never thwarted of his legitimate reward by them.

I have been a seeker of trout from my boyhood, and on all the expeditions in which this fish has been the ostensible purpose I have brought home more game than my creel showed. In fact, in my mature years I find I got more of nature into me, more of the woods, the wild, nearer to bird and beast, while threading my native streams for trout than in almost any other way. It furnished a good excuse to go forth; it pitched one in the right key; it sent one through the fat and marrowy places of field and wood. Then the fisherman has a harmless, preoccupied look; he is a kind of vagrant that nothing fears. He blends himself with the trees and the shadows. All his approaches are gentle and indirect. He times himself to the meandering, soliloquizing stream; its impulse bears him along. At the foot of the waterfall he sits sequestered and hidden in its volume of sound. The birds know he has no designs upon them, and the animals see that his mind is in the creek. His enthusiasm anneals him, and makes him pliable to the scenes and influences he moves among.

Then what acquaintance he makes with the stream ! He addresses himself to it as a lover to his mistress; he wooes it and stays with it till he knows its most hidden secrets. It runs through his thoughts not less than through its banks there; he feels the fret and thrust of every bar and boulder. Where it deepens, his purpose deepens; where it is shallow, he is indifferent. He knows how to interpret its every glance and dimple; its beauty haunts him for days.

I am sure I run no risk of overpraising the charm and attractiveness of a well-fed trout stream, every drop of water in it as bright and pure as if the

nymphs had brought it all the way from its source
in crystal goblets, and as cool as if it had been
hatched beneath a glacier. When the heated and
soiled and jaded refugee from the city first sees one,
he feels as if he would like to turn it into his bosom
and let it flow through him a few hours, it suggests
such healing freshness and newness. How his roily
thoughts would run clear; how the sediment would
go downstream! Could he ever have an impure or
an unwholesome wish afterward? The next best
thing he can do is to tramp along its banks and
surrender himself to its influence. If he reads it
intently enough, he will, in a measure, be taking it
into his mind and heart, and experiencing its salu-
tary ministrations.

Trout streams coursed through every valley my
boyhood knew. I crossed them, and was often lured
and detained by them, on my way to and from
school. We bathed in them during the long summer noons, and felt for the trout under their banks.
A holiday was a holiday indeed that brought per-
mission to go fishing over on Rose's Brook, or up
Hardscrabble, or in Meeker's Hollow; all-day trips,
from morning till night, through meadows and pastures and beechen woods, wherever the shy, limpid
stream led. What an appetite it developed! a hunger that was fierce and aboriginal, and that the wild
strawberries we plucked as we crossed the hill teased
rather than allayed. When but a few hours could

be had, gained perhaps by doing some piece of work about the farm or garden in half the allotted time, the little creek that headed in the paternal domain was handy; when half a day was at one's disposal, there were the hemlocks, less than a mile distant, with their loitering, meditative, log-impeded stream and their dusky, fragrant depths. Alert and wide-eyed, one picked his way along, startled now and then by the sudden bursting-up of the partridge, or by the whistling wings of the "dropping snipe," pressing through the brush and the briers, or finding an easy passage over the trunk of a prostrate tree, carefully letting his hook down through some tangle into a still pool, or standing in some high, sombre avenue and watching his line float in and out amid the moss-covered boulders. In my first essayings I used to go to the edge of these hemlocks, seldom dipping into them beyond the first pool where the stream swept under the roots of two large trees. From this point I could look back into the sunlit fields where the cattle were grazing; beyond, all was gloom and mystery; the trout were black, and to my young imagination the silence and the shadows were blacker. But gradually I yielded to the fascination and penetrated the woods farther and farther on each expedition, till the heart of the mystery was fairly plucked out. During the second or third year of my piscatorial experience I went through them, and through the pasture and meadow beyond, and through another strip of hemlocks, to where the little stream joined the main creek of the valley.

In June, when my trout fever ran pretty high, and an auspicious day arrived, I would make a trip to a stream a couple of miles distant, that came down out of a comparatively new settlement. It was a rapid mountain brook presenting many difficult problems to the young angler, but a very enticing stream for all that, with its two saw-mill dams, its pretty cascades, its high, shelving rocks sheltering the mossy nests of the phœbe-bird, and its general wild and forbidding aspects.

But a meadow brook was always a favorite. The trout like meadows; doubtless their food is more abundant there, and, usually, the good hiding-places are more numerous. As soon as you strike a meadow the character of the creek changes: it goes slower and lies deeper; it tarries to enjoy the high, cool banks and to half hide beneath them; it loves the willows, or rather the willows love it and shelter it from the sun; its spring runs are kept cool by the overhanging grass, and the heavy turf that faces its open banks is not cut away by the sharp hoofs of the grazing cattle. Then there are the bobolinks and the starlings and the meadowlarks, always interested spectators of the angler; there are also the marsh marigolds, the buttercups, or the spotted lilies, and the good angler is always an interested spectator of them. In fact, the patches of meadow land that lie in the angler's course are like the happy experiences in his own life, or like the fine passages in the poem he is reading; the pasture oftener contains the shallow and monotonous places. In the small streams the cattle scare the fish, and soil their element and break down their retreats under the banks. Woodland alternates the best with meadow: the creek loves to burrow under the roots of a great tree, to scoop out a pool after leaping over the prostrate trunk of one, and to pause at the foot of a ledge of moss-covered rocks, with ice-cold water dripping down. How straight the current goes for the rock! Note its corrugated, muscular appearance; it strikes and glances off, but accumulates, deepens with well-defined eddies above and to one side; on the edge of these the trout lurk and spring upon their prey.

The angler learns that it is generally some obstacle or hindrance that makes a deep place in the creek, as in a brave life; and his ideal brook is one that lies in deep, well-defined banks, yet makes many a shift from right to left, meets with many rebuffs and adventures, hurled back upon itself by rocks, waylaid by snags and trees, tripped up by precipices, but sooner or later reposing under meadow banks, deepening and eddying beneath bridges, or prosperous and strong in some level stretch of cultivated land with great elms shading it here and there.

But I early learned that from almost any stream

in a trout country the true angler could take trout, and that the great secret was this, that, whatever bait you used, worm, grasshopper, grub, or fly, there was one thing you must always put upon your hook, namely, your heart: when you bait your hook with your heart the fish always bite; they will jump clear from the water after it; they will dispute with each other over it; it is a morsel they love above everything else. With such bait I have seen the born angler (my grandfather was one) take a noble string of trout from the most unpromising waters, and on the most unpromising day. He used his hook so coyly and tenderly, he approached the fish with such address and insinuation, he divined the exact spot where they lay: if they were not eager, he humored them and seemed to steal by them; if they were playful and coquettish, he would suit his mood to theirs; if they were frank and sincere, he met them halfway; he was so patient and considerate, so entirely devoted to pleasing the critical trout, and so successful in his efforts, — surely his heart was upon his hook, and it was a tender, unctuous heart, too, as that of every angler is. How nicely he would measure the distance! how dexterously he would avoid an overhanging limb or bush and drop the line exactly in the right spot! Of course there was a pulse of feeling and sympathy to the extremity of that line. If your heart is a stone, however, or an empty husk, there is no use to put it upon your hook; it will not tempt the fish; the bait must be quick and fresh. Indeed, a certain quality of youth is indispensable to the successful angler, a certain unworldliness and readiness to invest yourself in an enterprise that does n't pay in the current coin. Not only is the angler, like the poet, born and not made, as Walton says, but there is a deal of the poet in him, and he is to be judged no more harshly; he is the victim of his genius: those wild streams, how they haunt him! he will play truant to dull care, and flee to them; their waters impart somewhat of their own perpetual youth to him. My grandfather when he was eighty years old would take down his pole as eagerly as any boy, and step off with wonderful elasticity toward the beloved streams; it used to try my young legs a good deal to follow him, specially on the return trip. And no poet was ever more innocent of worldly success or ambition. For, to paraphrase Tennyson, —

" Lusty trout to him were scrip and share,
And babbling waters more than cent for cent. "

He laid up treasures, but they were not in this world. In fact, though the kindest of husbands, I fear he was not what the country people call a " good provider," except in providing trout in their season, though it is doubtful if there was always fat in the house to fry them in. But he could tell you they were worse off than that at Valley Forge, and that trout, or any other fish, were good roasted in the ashes under the coals. He had the Walton requisite of loving quietness and contemplation, and was devout withal. Indeed, in many ways he was akin to those Galilee fishermen who were called to be fishers of men. How he read the Book and pored over it, even at times, I suspect, nodding over it, and laying it down only to take up his rod, over which, unless the trout were very dilatory and the journey very fatiguing, he never nodded!

II

The Delaware is one of our minor rivers, but it is a stream beloved of the trout. Nearly all its remote branches head in mountain springs, and its collected waters, even when warmed by the summer sun, are as sweet and wholesome as dew swept from the grass. The Hudson wins from it two streams that are fathered by the mountains from whose loins most of its beginnings issue, namely, the Rondout and the Esopus. These swell a more illustrious current than the Delaware, but the Rondout, one of the finest trout streams in the world, makes an uncanny alliance before it reaches its destination, namely, with the malarious Wallkill.

In the same nest of mountains from which they start are born the Neversink and the Beaverkill, streams of wondrous beauty that flow south and west into the Delaware. From my native hills I could catch glimpses of the mountains in whose laps these creeks were cradled, but it was not till after many years, and after dwelling in a country where trout are not found, that I returned to pay my respects to them as an angler.

My first acquaintance with the Neversink was made in company with some friends in 1869. We passed up the valley of the Big Ingin, marveling at its copious ice-cold springs, and its immense sweep of heavy-timbered mountain-sides. Crossing the range at its head, we struck the Neversink quite unexpectedly about the middle of the afternoon, at a point where it was a good-sized trout stream. It proved to be one of those black mountain brooks born of innumerable ice-cold springs, nourished in the shade, and shod, as it were, with thick-matted moss, that every camper-out remembers. The fish are as black as the stream and very wild. They dart from beneath the fringed rocks, or dive with the hook into the dusky depths, — an integral part of the silence and the shadows. The spell of the moss is over all. The fisherman's tread is noiseless, as he leaps from stone to stone and from ledge to ledge along the bed of the stream. How cool it is! He looks up the dark, silent defile, hears the solitary voice of the water, sees the decayed trunks of fallen trees bridging the stream, and all he has dreamed, when a boy, of the haunts of beasts of prey — the crouching feline tribes, especially if it be near nightfall and the gloom already deepening in the woods — comes freshly to mind, and he presses on, wary and alert, and speaking to his companions in low tones.

After an hour or so the trout became less abundant, and with nearly a hundred of the black sprites in our baskets we turned back. Here and there I saw the abandoned nests of the pigeons, sometimes half a dozen in one tree. In a yellow birch which the floods had uprooted, a number of nests were still in place, little shelves or platforms of twigs loosely arranged, and affording little or no protection to the eggs or the young birds against inclement weather.

Before we had reached our companions the rain set in again and forced us to take shelter under a balsam. When it slackened we moved on and soon came up with Aaron, who had caught his first trout, and, considerably drenched, was making his way toward camp, which one of the party had gone forward to build. After traveling less than a mile, we saw a smoke struggling up through the dripping trees, and in a few moments were all standing round a blazing fire. But the rain now commenced again, and fairly poured down through the trees, rendering the prospect of cooking and eating our supper there in the woods, and of passing the night on the ground without tent or cover of any kind, rather disheartening. We had been told of a bark shanty a couple of miles farther down the creek, and thitherward we speedily took up our line of march. When we were on the point of discontinuing the search, thinking we had been misinformed or had passed it by, we came in sight of a bark-peeling, in the midst of which a small log house lifted its naked rafters toward the now breaking sky. It had neither floor nor roof, and was less inviting on first sight than the open woods. But a board partition was still standing, out of which we built a rude porch on the east side of the house, large enough for us all to sleep under if well packed, and eat under if we stood up. There was plenty of well-seasoned timber lying about, and a fire was soon burning in front of our quarters that made the scene social and picturesque, especially when the frying-pans were brought into requisition, and the coffee, in charge of Aaron, who was an artist in this line, mingled its aroma with the wild-wood air. At dusk a balsam was felled, and the tips of the branches used to make a bed, which was more fragrant than soft; hemlock is better, because its needles are finer and its branches more elastic.

There was a spirt or two of rain during the night, but not enough to find out the leaks in our roof. It took the shower or series of showers of the next day to do that. They commenced about two o'clock in the afternoon. The forenoon had been fine, and we had brought into camp nearly three hundred trout;

but before they were half dressed, or the first pan-
fuls fried, the rain set in. First came short, sharp
dashes, then a gleam of treacherous sunshine, followed by more and heavier dashes. The wind was
in the southwest, and to rain seemed the easiest
thing in the world. From fitful dashes to a steady
pour the transition was natural. We stood huddled
together, stark and grim, under our cover, like hens
under a cart. The fire fought bravely for a time,
and retaliated with sparks and spiteful tongues of
flame; but gradually its spirit was broken, only a
heavy body of coal and half-consumed logs in the
centre holding out against all odds. The simmer-
ing fish were soon floating about in a yellow liquid
that did not look in the least appetizing. Point after
point gave way in our cover, till standing between
the drops was no longer possible. The water coursed
down the underside of the boards, and dripped in
our necks and formed puddles on our hat-brims.
We shifted our guns and traps and viands, till there
was no longer any choice of position, when the loaves
and the fishes, the salt and the sugar, the pork and
the butter, shared the same watery fate. The fire
was gasping its last. Little rivulets coursed about
it, and bore away the quenched but steaming coals
on their bosoms. The spring run in the rear of our
camp swelled so rapidly that part of the trout that
had been hastily left lying on its banks again found
themselves quite at home. For over two hours the

floods came down. About four o'clock Orville, who had not yet come from the day's sport, appeared. To say Orville was wet is not much; he was better than that, — he had been washed and rinsed in at least half a dozen waters, and the trout that he bore dangling at the end of a string hardly knew that they had been out of their proper element.

But he brought welcome news. He had been two or three miles down the creek, and had seen a log building, — whether house or stable he did not know, but it had the appearance of having a good roof, which was inducement enough for us instantly to leave our present quarters. Our course lay along an old wood-road, and much of the time we were to our knees in water. The woods were literally flooded everywhere. Every little rill and springlet ran like a mill-tail, while the main stream rushed and roared, foaming, leaping, lashing, its volume increased fifty-fold. The water was not roily, but of a rich coffee-color, from the leachings of the woods. No more trout for the next three days! we thought, as we looked upon the rampant stream.

After we had labored and floundered along for about an hour, the road turned to the left, and in a little stumpy clearing near the creek a gable uprose on our view. It did not prove to be just such a place as poets love to contemplate. It required a greater effort of the imagination than any of us were then capable of to believe it had ever been a

favorite resort of wood-nymphs or sylvan deities.
It savored rather of the equine and the bovine.
The bark-men had kept their teams there, horses
on the one side and oxen on the other, and no Her-
cules had ever done duty in cleansing the stables.
But there was a dry loft overhead with some straw,
where we might get some sleep, in spite of the rain
and the midges; a double layer of boards, standing
at a very acute angle, would keep off the former,
while the mingled refuse hay and muck beneath
would nurse a smoke that would prove a thorough
protection against the latter. And then, when Jim,
the two-handed, mounting the trunk of a prostrate
maple near by, had severed it thrice with easy and
familiar stroke, and, rolling the logs in front of the
shanty, had kindled a fire, which, getting the better
of the dampness, soon cast a bright glow over all,
shedding warmth and light even into the dingy
stable, I consented to unsling my knapsack and
accept the situation. The rain had ceased, and the
sun shone out behind the woods. We had trout
sufficient for present needs; and after my first meal
in an ox-stall, I strolled out on the rude log bridge
to watch the angry Neversink rush by. Its waters
fell quite as rapidly as they rose, and before sundown it looked as if we might have fishing again
on the morrow. We had better sleep that night
than either night before, though there were two
disturbing causes, — the smoke in the early part of

it, and the cold in the latter. The "no-see-ems" left in disgust ; and, though disgusted myself, I swallowed the smoke as best I could, and hugged my pallet of straw the closer. But the day dawned bright, and a plunge in the Neversink set me all right again. The creek, to our surprise and gratification, was only a little higher than before the rain, and some of the finest trout we had yet seen we caught that morning near camp.

We tarried yet another day and night at the old stable, but taking our meals outside squatted on the ground, which had now become quite dry. Part of the day I spent strolling about the woods, looking up old acquaintances among the birds, and, as always, half expectant of making some new ones. Curiously enough, the most abundant species were among those I had found rare in most other localities, namely, the small water-wagtail, the mourning ground warbler, and the yellow-bellied woodpecker. The latter seems to be the prevailing woodpecker through the woods of this region.

That night the midges, those motes that sting, held high carnival. We learned afterward, in the settlement below and from the barkpeelers, that it was the worst night ever experienced in that valley. We had done no fishing during the day, but had anticipated some fine sport about sundown. Accordingly Aaron and I started off between six and seven o'clock, one going upstream and the other down. The scene was charming. The sun shot up great spokes of light from behind the woods, and beauty, like a presence, pervaded the atmosphere. But torment, multiplied as the sands of the sea-shore, lurked in every tangle and thicket. In a thoughtless moment I removed my shoes and socks, and waded in the water to secure a fine trout that had accidentally slipped from my string and was helplessly floating with the current. This caused some delay and gave the gnats time to accumulate. Before I had got one foot half dressed I was enveloped in a black mist that settled upon my hands and neck and face, filling my ears with infinitesimal pipings and covering my flesh with infinitesimal bitings. I thought I should have to flee to the friendly fumes of the old stable, with " one stocking off and one stocking on;" but I got my shoe on at last, though not without many amusing interruptions and digressions.

In a few moments after this adventure I was in rapid retreat toward camp. Just as I reached the path leading from the shanty to the creek, my companion in the same ignoble flight reached it also, his hat broken and rumpled, and his sanguine countenance looking more sanguinary than I had ever before seen it, and his speech, also, in the highest degree inflammatory. His face and forehead were as blotched and swollen as if he had just run his head into a hornets' nest, and his manner as precipitate as if the whole swarm was still at his back.

No smoke or smudge which we ourselves could endure was sufficient in the earlier part of that evening to prevent serious annoyance from the same cause; but later a respite was granted us.

About ten o'clock, as we stood round our camp-fire, we were startled by a brief but striking display of the aurora borealis. My imagination had already been excited by talk of legends and of weird shapes and appearances, and when, on looking up toward the sky, I saw those pale, phantasmal waves of magnetic light chasing each other across the little opening above our heads, and at first sight seeming barely to clear the treetops, I was as vividly impressed as if I had caught a glimpse of a veritable spectre of the Neversink. The sky shook and trembled like a great white curtain.

After we had climbed to our loft and had lain down to sleep, another adventure befell us. This time a new and uninviting customer appeared upon the scene, the *genius loci* of the old stable, namely, the "fretful porcupine." We had seen the marks and work of these animals about the shanty, and had been careful each night to hang our traps, guns, etc., beyond their reach, but of the prickly night-walker himself we feared we should not get a view.

We had lain down some half hour, and I was just

on the threshold of sleep, ready, as it were, to pass through the open door into the land of dreams, when I heard outside somewhere that curious sound, — a sound which I had heard every night I spent in these woods, not only on this but on former expeditions, and which I had settled in my mind as proceeding from the porcupine, since I knew the sounds our other common animals were likely to make, — a sound that might be either a gnawing on some hard, dry substance, or a grating of teeth, or a shrill grunting.

Orville heard it also, and, raising up on his elbow, asked, " What is that ? "

" What the hunters call a ' porcupig,' " said I.

" Sure ? "

" Entirely so."

" Why does he make that noise ? "

" It is a way he has of cursing our fire," I replied. " I heard him last night also."

" Where do you suppose he is ? " inquired my companion, showing a disposition to look him up.

" Not far off, perhaps fifteen or twenty yards from our fire, where the shadows begin to deepen."

Orville slipped into his trousers, felt for my gun, and in a moment had disappeared down through the scuttle hole. I had no disposition to follow him, but was rather annoyed than otherwise at the disturbance. Getting the direction of the sound, he went picking his way over the rough, uneven ground, and, when he got where the light failed him, poking every doubtful object with the end of his gun. Presently he poked a light grayish object, like a large round stone, which surprised him by moving off. On this hint he fired, making an incurable wound in the " porcupig," which, nevertheless, tried harder than ever to escape. I lay listening, when, close on the heels of the report of the gun, came excited shouts for a revolver. Snatching up my Smith and Wesson, I hastened, shoeless and hatless, to the scene of action, wondering what was up. I found my companion struggling to detain, with the end of the gun, an uncertain object that was trying to crawl off into the darkness. "Look out!" said Orville, as he saw my bare feet, " the quills are lying thick around here."

And so they were; he had blown or beaten them nearly all off the poor creature's back, and was in a fair way completely to disable my gun, the ramrod of which was already broken and splintered clubbing his victim. But a couple of shots from the revolver, sighted by a lighted match, at the head of the animal, quickly settled him.

He proved to be an unusually large Canada porcupine, — an old patriarch, gray and venerable, with spines three inches long, and weighing, I should say, twenty pounds. The build of this animal is much like that of the woodchuck, that is, heavy and pouchy. The nose is blunter than that of the woodchuck, the limbs stronger, and the tail broader and heavier. Indeed, the latter appendage is quite club-like, and the animal can, no doubt, deal a smart blow with it. An old hunter with whom I talked thought it aided them in climbing. They are inveterate gnawers, and spend much of their time in trees gnawing the bark. In winter one will take up its abode in a hemlock, and continue there till the tree is quite denuded. The carcass emitted a peculiar, offensive odor, and, though very fat, was not in the least inviting as game. If it is part of the economy of nature for one animal to prey upon some other beneath it, then the poor devil has indeed a mouthful that makes a meal off the porcupine. Panthers and lynxes have essayed it, but have invariably left off at the first course, and have afterwards been found dead, or nearly so, with their heads puffed up like a pincushion, and the quills protruding on all sides. A dog that understands the business will manœuvre round the porcupine till he gets an opportunity to throw it over on its back, when he fastens on its quilless underbody. Aaron was puzzled to know how long-parted friends could embrace, when it was suggested that the quills could be depressed or elevated at pleasure.

The next morning boded rain; but we had become thoroughly sated with the delights of our present quarters, outside and in, and packed up our traps to leave. Before we had reached the clearing, three miles below, the rain set in, keeping up a lazy monotonous drizzle till the afternoon.

The clearing was quite a recent one, made mostly by barkpeelers, who followed their calling in the mountains round about in summer, and worked in their shops making shingle in winter. The Biscuit Brook came in here from the west, — a fine, rapid trout stream six or eight miles in length, with plenty of deer in the mountains about its head. On its banks we found the house of an old woodman, to whom we had been directed for information about the section we proposed to traverse.

" Is the way very difficult," we inquired, " across from the Neversink into the head of the Beaver-kill ? "

" Not to me; I could go it the darkest night ever was. And I can direct you so you can find the way without any trouble. You go down the Neversink about a mile, when you come to Highfall Brook, the first stream that comes down on the right. Follow up it to Jim Reed's shanty, about three miles. Then cross the stream, and on the left bank, pretty well up on the side of the mountain, you will find a wood-road, which was made by a fellow below here who stole some ash logs off the top of the ridge last winter and drew them out on the snow. When the road first begins to tilt over the mountain, strike down to your left, and you can reach the Beaverkill before sundown."

SPECKLED TROUT

As it was then after two o'clock, and as the distance was six or eight of these terrible hunters' miles, we concluded to take a whole day to it, and wait till next morning. The Beaverkill flowed west, the Neversink south, and I had a mortal dread of getting entangled amid the mountains and valleys that lie in either angle.

Besides, I was glad of another and final opportunity to pay my respects to the finny tribes of the Neversink. At this point it was one of the finest trout streams I had ever beheld. It was so sparkling, its bed so free from sediment or impurities of any kind, that it had a new look, as if it had just come from the hand of its Creator. I tramped along its margin upward of a mile that afternoon, part of the time wading to my knees, and casting my hook, baited only with a trout's fin, to the opposite bank. Trout are real cannibals, and make no bones, and break none either, in lunching on each other. A friend of mine had several in his spring, when one day a large female trout gulped down one of her male friends, nearly one third her own size, and went around for two days with the tail of her liege lord protruding from her mouth! A fish's eye will do for bait, though the anal fin is better. One of the natives here told me that when he wished to catch large trout (and I judged he never fished for any other, — I never do), he used for bait the bullhead, or dart, a little fish an inch and a half or two inches long, that rests on the pebbles near shore and darts quickly, when disturbed, from point to point. " Put that on your hook," said he, " and if there is a big fish in the creek, he is bound to have it." But the darts were not easily found ; the big fish, I concluded, had cleaned them all out; and, then, it was easy enough to supply our wants with a fin.

Declining the hospitable offers of the settlers, we spread our blankets that night in a dilapidated shingle-shop on the banks of the Biscuit Brook, first flooring the damp ground with the new shingle that lay piled in one corner. The place had a great-throated chimney with a tremendous expanse of fireplace within, that cried " More!" at every morsel of wood we gave it.

But I must hasten over this part of the ground, nor let the delicious flavor of the milk we had that morning for breakfast, and that was so delectable after four days of fish, linger on my tongue; nor yet tarry to set down the talk of that honest, weather-worn passer-by who paused before our door, and every moment on the point of resuming his way, yet stood for an hour and recited his adventures hunting deer and bears on these mountains. Having replenished our stock of bread and salt pork at the house of one of the settlers, midday found us at Reed's shanty, — one of those temporary structures erected by the bark jobber to lodge and board his " hands " near their work. Jim not being at home, we could gain no information from the "women folks" about the way, nor from the men who had just come in to dinner; so we pushed on, as near as we could, according to the instructions we had previously received. Crossing the creek, we forced our way up the side of the mountain, through a perfect *cheval-de-frise* of fallen and peeled hemlocks, and, entering the dense woods above, began to look anxiously about for the wood-road. My companions at first could see no trace of it; but knowing that a casual wood-road cut in winter, when there was likely to be two or three feet of snow on the ground, would present only the slightest indications to the eye in summer, I looked a little closer, and could make out a mark or two here and there. The larger trees had been avoided, and the axe used only on the small saplings and underbrush, which had been lopped off a couple of feet from the ground. By being constantly on the alert, we followed it till near the top of the mountain; but, when looking to see it " tilt " over the other side, it disappeared altogether. Some stumps of the black cherry were found, and a solitary pair of snow-shoes was hanging high and dry on a branch, but no further trace of human hands could we see. While we were resting here a couple of hermit thrushes, one of them with some sad defect in his vocal powers which barred him from uttering more than a few notes of his song, gave voice to the solitude of the place. This was the second instance in which I have observed a song-bird with apparently some organic defect in its instrument. The other case was that of a bobolink, which, hover in mid-air and inflate its throat as it might, could only force out a few incoherent notes. But the bird in each case presented this striking contrast to human examples of the kind, that it was apparently just as proud of itself, and just as well satisfied with its performance, as were its more successful rivals.

After deliberating some time over a pocket compass which I carried, we decided upon our course, and held on to the west. The descent was very gradual. Traces of bear and deer were noted at different points, but not a live animal was seen.

About four o'clock we reached the bank of a stream flowing west. Hail to the Beaverkill ! and we pushed on along its banks. The trout were plenty, and rose quickly to the hook ; but we held on our way, designing to go into camp about six o'clock. Many inviting places, first on one bank, then on the other, made us linger, till finally we reached a smooth, dry place overshadowed by balsam and hemlock, where the creek bent around a little flat, which was so entirely to our fancy that we unslung our knapsacks at once. While my companions were cutting wood and making other preparations for the night, it fell to my lot, as the most successful angler, to provide the trout for supper and breakfast. How shall I describe that wild beautiful stream, with features so like those of all other mountain streams? And yet, as I saw it in the deep twilight of those woods on that June afternoon, with its steady, even flow, and its tranquil, many-voiced murmur, it made an impression upon my mind distinct and peculiar, fraught in an eminent degree with the charm of seclusion and remoteness. The solitude was perfect, and I felt that strangeness and insignificance which the civilized man must always feel when opposing himself to such a vast scene of silence and wildness. The trout were quite black, like all wood trout, and took the bait eagerly. I followed the stream till the deepening shadows warned me to turn back. As I neared camp, the fire shone far through the trees, dispelling the gathering gloom, but blinding my eyes to all obstacles at my feet. I was seriously disturbed on arriving to find that one of my companions had cut an ugly gash in his shin with the axe while felling a tree. As we did not carry a fifth wheel, it was not just the time or place to have any of our members crippled, and I had bodings of evil. But, thanks to the healing virtues of the balsam which must have adhered to the blade of the axe, and double thanks to the court-plaster with which Orville had supplied himself before leaving home, the wounded leg, by being favored that night and the next day, gave us little trouble.

That night we had our first fair and square camping out, — that is, sleeping on the ground with no shelter over us but the trees, — and it was in many respects the pleasantest night we spent in the woods. The weather was perfect and the place was perfect, and for the first time we were exempt from the midges and smoke; and then we appreciated the clean new page we had to work on. Nothing is so acceptable to the camper-out as a pure article in the way of woods and waters. Any admixture of human relics mars the spirit of the scene. Yet I am willing to confess that, before we were through those woods, the marks of an axe in a tree were a welcome sight. On resuming our march next day we followed the right bank of the Beaverkill, in order to strike a stream which flowed in from the north, and which was the outlet of Balsam Lake, the objective point of that day's march. The distance to the lake from our camp could not have been over six or seven miles; yet, traveling as we did, without path or guide, climbing up banks, plunging into ravines, making detours around swampy places, and forcing our way through woods choked up with much fallen and decayed timber, it seemed at least twice that distance, and the mid-afternoon sun was shining when we emerged into what is called the " Quaker Clearing," ground that I had been over nine years before, and that lies about two miles south of the lake. From this point we had a well-worn path that led us up a sharp rise of ground, then through level woods till we saw the bright gleam of the water through the trees.

I am always struck, on approaching these little mountain lakes, with the extensive preparation that is made for them in the conformation of the ground. I am thinking of a depression, or natural basin, in the side of the mountain or on its top, the brink of which I shall reach after a little steep climbing ; but instead of that, after I have accomplished the ascent, I find a broad sweep of level or gently undulating woodland that brings me after a half hour or so to the lake, which lies in this vast lap like a drop of water in the palm of a man's hand.

Balsam Lake was oval-shaped, scarcely more than half a mile long and a quarter of a mile wide, but presented a charming picture, with a group of dark gray hemlocks filling the valley about its head, and the mountains rising above and beyond. We found a bough house in good repair, also a dug-out and paddle and several floats of logs. In the dug-out I was soon creeping along the shady side of the lake, where the trout were incessantly jumping for a species of black fly, that, sheltered from the slight breeze, were dancing in swarms just above the surface of the water. The gnats were there in swarms also, and did their best toward balancing the accounts by preying upon me while I preyed upon the trout which preyed upon the flies. But by dint of keeping my hands, face, and neck constantly wet, I am convinced that the balance of blood was on my side. The trout jumped most within a foot or two of shore, where the water was only a few inches deep. The shallowness of the water, perhaps, accounted for the inability of the fish to do more than lift their heads above the surface. They came up mouths wide open, and dropped back again in the most impotent manner. Where there is any depth of water, a trout will jump several feet into the air; and where there is a solid, unbroken sheet or column, they will scale falls and dams fifteen feet high.

We had the very cream and flower of our trout-fishing at this lake. For the first time we could use the fly to advantage; and then the contrast between laborious tramping along shore, on the one hand, and sitting in one end of a dug-out and casting your line right and left with no fear of entanglement in brush or branch, while you were gently propelled along, on the other, was of the most pleasing character.

There were two varieties of trout in the lake, — what it seems proper to call silver trout and golden trout; the former were the slimmer, and seemed to keep apart from the latter. Starting from the outlet and working round on the eastern side toward the head, we invariably caught these first. They glanced in the sun like bars of silver. Their sides and bellies were indeed as white as new silver. As we neared the head, and especially as we came near

a space occupied by some kind of watergrass that
grew in the deeper part of the lake, the other variety
would begin to take the hook, their bellies a bright
gold color, which became a deep orange on their
fins; and as we returned to the place of departure
with the bottom of the boat strewn with these bright
forms intermingled, it was a sight not soon to be
forgotten. It pleased my eye so, that I would fain
linger over them, arranging them in rows and studying the various hues and tints. They were of nearly
a uniform size, rarely one over ten or under eight
inches in length, and it seemed as if the hues of all
the precious metals and stones were reflected from
their sides. The flesh was deep salmon-color; that
of brook trout is generally much lighter. Some
hunters and fishers from the valley of the Mill Brook,
whom we met here, told us the trout were much
larger in the lake, though far less numerous than
they used to be. Brook trout do not grow large till
they become scarce. It is only in streams that have
been long and much fished that I have caught them
as much as sixteen inches in length.

The " porcupigs " were numerous about the lake,
and not at all shy. One night the heat became so
intolerable in our oven-shaped bough house that I
was obliged to withdraw from under its cover and
lie down a little to one side. Just at daybreak, as
I lay rolled in my blanket, something awoke me.
Lifting up my head, there was a porcupine with his

forepaws on my hips. He was apparently as much surprised as I was; and to my inquiry as to what he at that moment might be looking for, he did not pause to reply, but hitting me a slap with his tail which left three or four quills in my blanket, he scampered off down the hill into the brush.

Being an observer of the birds, of course every curious incident connected with them fell under my notice. Hence, as we stood about our camp-fire one afternoon looking out over the lake, I was the only one to see a little commotion in the water, half hidden by the near branches, as of some tiny swimmer struggling to reach the shore. Rushing to its rescue in the canoe, I found a yellow-rumped warbler, quite exhausted, clinging to a twig that hung down into the water. I brought the drenched and helpless thing to camp, and, putting it into a basket, hung it up to dry. An hour or two afterward I heard it fluttering in its prison, and cautiously lifted the lid to get a better glimpse of the lucky captive, when it darted out and was gone in a twinkling. How came it in the water? That was my wonder, and I can only guess that it was a young bird that had never before flown over a pond of water, and, seeing the clouds and blue sky so perfect down there, thought it was a vast opening or gateway into another summer land, perhaps a short cut to the tropics, and so got itself into trouble. How my eye was delighted also with the redbird that alighted for a moment on a dry branch above the lake, just where a ray of light from the setting sun fell full upon it! A mere crimson point, and yet how it offset that dark, sombre background!

I have thus run over some of the features of an ordinary trouting excursion to the woods. People inexperienced in such matters, sitting in their rooms and thinking of these things, of all the poets have sung and romancers written, are apt to get sadly taken in when they attempt to realize their dreams. They expect to enter a sylvan paradise of trout, cool retreats, laughing brooks, picturesque views, and balsamic couches, instead of which they find hunger, rain, smoke, toil, gnats, mosquitoes, dirt, broken rest, vulgar guides, and salt pork; and they are very apt not to see where the fun comes in. But he who goes in a right spirit will not be disappointed, and will find the taste of this kind of life better, though bitterer, than the writers have described.

AN IDYL OF THE HONEY-BEE

*At Woodchuck Lodge on his Old Homestead
in Roxbury, New York*

EDITOR'S NOTE

In essays like "Pastoral Bees" and "An Idyl of the Honey-Bee," Burroughs writes about bees with an ease that reflects a familiarity stemming back to his days as an observant farm boy. He later kept bees at his Hudson Valley farm and sometimes said that if he had not been a writer he might have been a full-time beekeeper. He compared himself to a bee in his love of catching the early whiffs of spring pollen. As a writer, he compared his method to that of a bee making honey. Like the bee adding secretions to sweet water and transforming it into honey, the writer draws his material from nature but adds something of himself to it in the transformative process of his imagination. "The bee is therefore the type of the true poet, the true artist," Burroughs writes. "It is this drop of herself that gives the delicious sting to her sweet." To Burroughs the sport of tracking down wild bees yields a taste of honey that contains not only thyme, linden, or sumac but also something from the experience itself. Bee hunting engages Burroughs as a participant in the natural world, and it shows what he continually

EDITOR'S NOTE

finds so attractive as a naturalist, a series of secrets, riddles, and mysteries that are immanent and natural. "An Idyl of the Honey-Bee" is reprinted from *Pepacton,* published in 1881.

F.B.

AN IDYL OF THE HONEY-BEE

THERE is no creature with which man has surrounded himself that seems so much like a product of civilization, so much like the result of development on special lines and in special fields, as the honey-bee. Indeed, a colony of bees, with their neatness and love of order, their division of labor, their public-spiritedness, their thrift, their complex economies, and their inordinate love of gain, seems as far removed from a condition of rude nature as does a walled city or a cathedral town. Our native bee, on the other hand, the "burly, dozing humblebee," affects one more like the rude, untutored savage. He has learned nothing from experience. He lives from hand to mouth. He luxuriates in time of plenty, and he starves in time of scarcity. He lives in a rude nest, or in a hole in the ground, and in small communities; he builds a few deep cells or sacks in which he stores a little honey and bee-bread for his young, but as a worker in wax he is of the most primitive and awkward. The Indian regarded the honey-bee as an ill omen. She was the white man's fly. In fact, she was the epitome of the white man himself. She has the white man's craftiness, his industry, his architectural skill, his neatness and love of system, his foresight; and, above all, his eager, miserly habits. The honey-bee's great ambition is to be rich, to lay up great stores, to possess the sweet of every flower that blooms. She is more than provident. Enough will not satisfy her; she must have all she can get by hook or by crook. She comes from the oldest country, Asia, and thrives best in the most fertile and long-settled lands.

Yet the fact remains that the honey-bee is essentially a wild creature, and never has been and cannot be thoroughly domesticated. Its proper home is the woods, and thither every new swarm counts on going; and thither many do go in spite of the care and watchfulness of the bee-keeper. If the woods in any given locality are deficient in trees with suitable cavities, the bees resort to all sorts of makeshifts; they go into chimneys, into barns and outhouses, under stones, into rocks, etc. Several chimneys in my locality with disused flues are taken possession of by colonies of bees nearly every season. One day, while bee-hunting, I developed a line that went toward a farmhouse where I had reason to believe no bees were kept. I followed it up and questioned the farmer about his bees. He said he kept no bees, but that a swarm had taken possession of his chimney, and another had gone under the clapboards in the gable end of his house. He had taken a large lot of honey out of both places the year before. Another farmer told me that one day his family had seen a number of bees examining a knothole in the side of his house; the next day, as they were sitting down to dinner, their attention was attracted by a loud humming noise, when they discovered a swarm of bees settling upon the side of the house and pouring into the knothole. In subsequent years other swarms came to the same place.

Apparently, every swarm of bees, before it leaves the parent hive, sends out exploring parties to look up the future home. The woods and groves are searched through and through, and no doubt the privacy of many a squirrel and many a wood-mouse is intruded upon. What cozy nooks and retreats they do spy out, so much more attractive than the painted hive in the garden, so much cooler in summer and so much warmer in winter!

The bee is in the main an honest citizen: she prefers legitimate to illegitimate business; she is never an outlaw until her proper sources of supply fail; she will not touch honey as long as honey-yielding flowers can be found; she always prefers to go to the fountain-head, and dislikes to take her sweets at second hand. But in the fall, after the flowers have failed, she can be tempted. The bee-hunter takes advantage of this fact; he betrays her with a little honey. He wants to steal her stores, and he first encourages her to steal his, then follows the thief home with her booty. This is the whole trick of the bee-hunter. The bees never suspect his game, else by taking a circuitous route they could easily baffle him. But the honey-bee has absolutely no wit or cunning outside of her special gifts as a gatherer and storer of honey. She is a simple-minded creature, and can be imposed upon by any novice. Yet it is not every novice that can find a bee-tree. The sportsman may track his game to its retreat by the aid of his dog, but in hunting the honey-bee one must be his own dog, and track his game through an element in which it leaves no trail. It is a task for a sharp, quick eye, and may test the resources of the best woodcraft. One autumn, when I devoted much time to this pursuit, as the best means of getting at nature and the open-air exhilaration, my eye became so trained that bees were nearly as easy to it as birds. I saw and heard bees wherever I went. One day, standing on a street corner in a great city, I saw above the trucks and the traffic a line of bees carrying off sweets from some grocery or confectionery shop.

One looks upon the woods with a new interest when he suspects they hold a colony of bees. What a pleasing secret it is, — a tree with a heart of comb honey, a decayed oak or maple with a bit of Sicily or Mount Hymettus stowed away in its trunk or branches; secret chambers where lies hidden the wealth of ten thousand little freebooters, great nuggets and wedges of precious ore gathered with risk and labor from every field and wood about!

But if you would know the delights of bee-hunting, and how many sweets such a trip yields besides honey, come with me some bright, warm, late September or early October day. It is the golden season of the year, and any errand or pursuit that takes us abroad upon the hills or by the painted woods and along the amber-colored streams at such a time is enough. So, with haversacks filled with grapes and peaches and apples and a bottle of milk, — for we shall not be home to dinner, — and armed with a compass, a hatchet, a pail, and a box with a piece of comb honey neatly fitted into it, — any box the size of your hand with a lid will do nearly as well as the elaborate and ingenious contrivance of the regular bee-hunter, — we sally forth. Our course at first lies along the highway under great chestnut-trees whose nuts are just dropping, then through an orchard and across a little creek, thence gently rising through a long series of cultivated fields toward some high uplying land behind which rises a rugged wooded ridge or mountain, the most sightly point in all this section. Behind this ridge for several miles the country is wild, wooded, and rocky, and is no doubt the home of many swarms of wild bees. What a gleeful uproar the robins, cedar-birds, high-holes, and cow blackbirds make amid the black cherry-trees as we pass along! The raccoons, too, have been here after black cherries, and we see their marks at various points. Several crows are walking about a newly sowed wheat-field we pass through, and we pause to note their graceful movements and glossy coats. I have seen no bird walk the ground with just the same air the crow does. It is not exactly pride; there is no strut or swagger in it, though perhaps just a little condescension; it is the contented, complaisant, and self-possessed gait of a lord over his domains. All these acres are mine, he says, and all these crops; men plow and sow for me, and I stay here or go there, and find life sweet and good wherever I am. The hawk looks awkward and out of place on the ground; the game-birds hurry and skulk; but the crow is at home, and treads the earth as if there were none to molest or make him afraid.

The crows we have always with us, but it is not every day or every season that one sees an eagle. Hence I must preserve the memory of one I saw the last day I went bee-hunting. As I was laboring up the side of a mountain at the head of a valley, the noble bird sprang from the top of a dry tree above me and came sailing directly over my head. I saw him bend his eye down upon me, and I could hear the low hum of his plumage as if the web of every quill in his great wings vibrated in his strong, level flight. I watched him as long as my eye could hold him. When he was fairly clear of the mountain, he began that sweeping spiral movement in which he climbs the sky. Up and up he went, without once breaking his majestic poise, till he appeared to sight some far-off alien geography, when he bent his course thitherward and gradually vanished in the blue depths. The eagle is a bird of large ideas; he embraces long distances; the continent is his home. I never look upon one without emotion; I follow him with my eye as long as I can. I think of Canada, of the Great Lakes, of the Rocky Mountains, of the wild and sounding seacoast. The waters are his, and the woods and the inaccessible cliffs. He pierces behind the veil of the storm, and his joy is height and depth and vast spaces.

We go out of our way to touch at a spring run in the edge of the woods, and are lucky to find a single scarlet lobelia lingering there. It seems almost to light up the gloom with its intense bit of color. Beside a ditch in a field beyond, we find the great blue lobelia, and near it, amid the weeds and wild grasses and purple asters, the most beautiful of our fall flowers, the fringed gentian. What a rare and delicate, almost aristocratic look the gentian has amid its coarse, unkempt surroundings! It does not lure the bee, but it lures and holds every passing human eye. If we strike through the corner of yonder woods, where the ground is moistened by hidden springs, and where there is a little opening amid the trees, we shall find the closed gentian, a rare flower in this locality. I had walked this way many times before I chanced upon its retreat, and then I was following a line of bees. I lost the bees, but I got the gentians. How curious this flower looks with its deep blue petals folded together so tightly, — a bud and yet a blossom! It is the nun among our wild flowers, — a form closely veiled and cloaked. The buccaneer bumblebee sometimes tries to rifle it of its sweets. I have seen the blossom with the bee entombed in it. He had forced his way into the virgin corolla as if determined to know its secret, but he had never returned with the knowledge he had gained.

After a refreshing walk of a couple of miles we reach a point where we will make our first trial, — a high stone wall that runs parallel with the wooded ridge referred to, and separated from it by a broad field. There are bees at work there on that goldenrod, and it requires but little manœuvring to sweep one into our box. Almost any other creature rudely and suddenly arrested in its career, and clapped into a cage in this way, would show great confusion and alarm. The bee is alarmed for a moment, but the bee has a passion stronger than its love of life or fear of death, namely, desire for honey, not simply to eat, but to carry home as booty. "Such rage of honey in their bosom beats," says Virgil. It is quick to catch the scent of honey in the box, and as quick to fall to filling itself. We now set the box down upon the wall and gently remove the cover. The bee is head and shoulders in one of the half-filled cells, and is oblivious to everything else about it. Come rack, come ruin, it will die at work. We step back a few paces, and sit down upon the ground so as to bring the box against the blue sky as a background. In two or three minutes the bee is seen rising slowly and heavily from the box. It seems loath to leave so much honey behind, and it marks the place well. It mounts aloft in a rapidly increasing spiral, surveying the near and minute objects first, then the larger and more distant, till, having circled above the spot five or six times and taken all its bearings, it darts away for home. It is a good eye that holds fast to the bee till it is fairly off. Sometimes one's head will swim following it, and often one's eyes are put out by the sun. This bee gradually drifts down the hill, then strikes off toward a farmhouse half a mile away where I know bees are kept. Then we try another and another, and the third bee, much to our satisfaction, goes straight toward the woods. We can see the brown speck against the darker background for many yards. The regular bee-hunter professes to be able to tell a wild bee from a tame one by the color, the former, he says, being lighter. But there is no difference; they are alike in color and in manner. Young bees are lighter than old, and that is all there is of it. If a bee lived many years in the woods, it would doubtless come to have some distinguishing marks, but the life of a bee is only a few months at the farthest, and no change is wrought in this brief time.

Our bees are all soon back, and more with them, for we have touched the box here and there with the cork of a bottle of anise oil, and this fragrant and pungent oil will attract bees half a mile or more. When no flowers can be found, this is the quickest way to obtain a bee.

It is a singular fact that when the bee first finds the hunter's box, its first feeling is one of anger; it is as mad as a hornet; its tone changes, it sounds its shrill war trumpet and darts to and fro, and gives vent to its rage and indignation in no uncertain manner. It seems to scent foul play at once. It says, "Here is robbery; here is the spoil of some hive, maybe my own," and its blood is up. But its ruling passion soon comes to the surface, its avarice gets the better of its indignation, and it seems to say, "Well, I had better take possession of this and carry it home." So after many feints and approaches and dartings off with a loud angry hum as if it would none of it, the bee settles down and fills itself.

It does not entirely cool off and get soberly to

work till it has made two or three trips home with
its booty. When other bees come, even if all from
the same swarm, they quarrel and dispute over the
box, and clip and dart at each other like bantam
cocks. Apparently the ill feeling which the sight
of the honey awakens is not one of jealousy or
rivalry, but wrath.

A bee will usually make three or four trips from
the hunter's box before it brings back a companion.
I suspect the bee does not tell its fellows what it
has found, but that they smell out the secret; it
doubtless bears some evidence with it upon its feet
or proboscis that it has been upon honeycomb and
not upon flowers, and its companions take the hint
and follow, arriving always many seconds behind.
Then the quantity and quality of the booty would
also betray it. No doubt, also, there are plenty of
gossips about a hive that note and tell everything.
"Oh, did you see that? Peggy Mel came in a few
moments ago in great haste, and one of the upstairs
packers says she was loaded till she groaned with
apple-blossom honey, which she deposited, and then
rushed off again like mad. Apple-blossom honey in
October! Fee, fi, fo, fum! I smell something! Let's
after."

In about half an hour we have three well-defined
lines of bees established, — two to farmhouses and
one to the woods, and our box is being rapidly
depleted of its honey. About every fourth bee goes

to the woods, and now that they have learned the way thoroughly, they do not make the long preliminary whirl above the box, but start directly from it. The woods are rough and dense and the hill steep, and we do not like to follow the line of bees until we have tried at least to settle the problem as to the distance they go into the woods, — whether the tree is on this side of the ridge or into the depth of the forest on the other side. So we shut up the box when it is full of bees and carry it about three hundred yards along the wall from which we are operating. When liberated, the bees, as they always will in such cases, go off in the same directions they have been going; they do not seem to know that they have been moved. But other bees have followed our scent, and it is not many minutes before a second line to the woods is established. This is called cross-lining the bees. The new line makes a sharp angle with the other line, and we know at once that the tree is only a few rods in the woods. The two lines we have established form two sides of a triangle, of which the wall is the base; at the apex of the triangle, or where the two lines meet in the woods, we are sure to find the tree. We quickly follow up these lines, and where they cross each other on the side of the hill we scan every tree closely. I pause at the foot of an oak and examine a hole near the root; now the bees are in this tree and their entrance is on the upper side near the ground not two feet from the hole I peer into, and yet so quiet and secret is their going and coming that I fail to discover them and pass on up the hill. Failing in this direction, I return to the oak again, and then perceive the bees going out in a small crack in the tree. The bees do not know they are found out and that the game is in our hands, and are as oblivious of our presence as if we were ants or crickets. The indications are that the swarm is a small one, and the store of honey trifling. In "taking up" a bee-tree it is usual first to kill or stupefy the bees with the fumes of burning sulphur or with tobacco smoke. But this course is impracticable on the present occasion, so we boldly and ruthlessly assault the tree with an axe we have procured. At the first blow the bees set up a loud buzzing, but we have no mercy, and the side of the cavity is soon cut away and the interior with its white-yellow mass of comb honey is exposed, and not a bee strikes a blow in defense of its all. This may seem singular, but it has nearly always been my experience. When a swarm of bees are thus rudely assaulted with an axe, they evidently think the end of the world has come, and, like true misers as they are, each one seizes as much of the treasure as it can hold; in other words, they all fall to and gorge themselves with honey, and calmly await the issue. While in this condition they make no defense, and will not sting unless taken hold of. In fact, they are as harmless as flies. Bees are always to be managed with boldness and decision. Any halfway measures, any timid poking about, any feeble attempts to reach their honey, are sure to be quickly resented. The popular notion that bees have a special antipathy toward certain persons and a liking for certain others has only this fact at the bottom of it: they will sting a person who is afraid of them and goes skulking and dodging about, and they will not sting a person who faces them boldly and has no dread of them. They are like dogs. The way to disarm a vicious dog is to show him you do not fear him; it is his turn to be afraid then. I never had any dread of bees, and am seldom stung by them. I have climbed up into a large chestnut that contained a swarm in one of its cavities and chopped them out with an axe, being obliged at times to pause and brush the bewildered bees from my hands and face, and not been stung once. I have chopped a swarm out of an apple-tree in June, and taken out the cards of honey and arranged them in a hive, and then dipped out the bees with a dipper, and taken the whole home with me in pretty good condition, with scarcely any opposition on the part of the bees. In reaching your hand into the cavity to detach and remove the comb you are pretty sure to get stung, for when you touch the "business end" of a bee, it will sting even though its head be off. But the bee carries the antidote to its own poison. The best remedy for bee sting is honey, and when your hands are besmeared with honey, as they are sure to be on such occasions, the wound is scarcely more painful than the prick of a pin. Assault your bee-tree, then, boldly with your axe, and you will find that when the honey is exposed every bee has surrendered, and the whole swarm is cowering in helpless bewilderment and terror. Our tree yields only a few pounds of honey, not enough to have lasted the swarm till January, but no matter: we have the less burden to carry.

In the afternoon we go nearly half a mile farther along the ridge to a corn-field that lies immediately in front of the highest point of the mountain. The view is superb; the ripe autumn landscape rolls away to the east, cut through by the great placid river; in the extreme north the wall of the Catskills stands out clear and strong, while in the south the mountains of the Highlands bound the view. The day is warm, and the bees are very busy there in that neglected corner of the field, rich in asters, fleabane, and goldenrod. The corn has been cut, and upon a stout but a few rods from the woods, which here drop quickly down from the precipitous heights, we set up our bee-box, touched again with the pungent oil. In a few moments a bee has found it; she comes up to leeward, following the scent. On leaving the box, she goes straight toward the woods. More bees quickly come, and it is not long before the line is well established. Now we have recourse to the same tactics we employed before, and move along the ridge to another field to get our cross-line. But the bees still go in almost the same direction they did from the corn stout. The tree is then either on the top of the mountain or on the other or west side of it. We hesitate to make the plunge into the woods and seek to scale those precipices, for the eye can plainly see what is before us. As the afternoon sun gets lower, the bees are seen with wonderful distinctness. They fly toward and under the sun, and are in a strong light, while the near woods which form the background are in deep shadow. They look like large luminous motes. Their swiftly vibrating, transparent wings surround their bodies with a shining nimbus that makes them visible for a long distance. They seem magnified many times. We see them bridge the little gulf between us and the woods, then rise up over the treetops with their burdens, swerving neither to the right hand nor to the left. It is almost pathetic to see them labor so, climbing the mountain and unwittingly guiding us to their treasures. When the sun gets down so that his direction corresponds exactly with the course of the bees, we make the plunge. It proves even harder climbing than we had anticipated; the mountain is faced by a broken and irregular wall of rock, up which we pull ourselves slowly and cautiously by main strength. In half an hour, the perspiration streaming from every pore, we reach the summit. The trees here are all small, a second growth, and we are soon convinced the bees are not here. Then down we go on the other side, clambering down the rocky stairways till we reach quite a broad plateau that forms something like the shoulder of the mountain. On the brink of this there are many large hemlocks, and we scan them closely and rap upon them with our axe. But not a bee is seen or heard; we do not seem as near the tree as we were in the fields below; yet, if some divinity would only whisper the fact to us, we are within a few rods of the coveted prize, which is not in one of the large hemlocks or oaks that absorb our attention, but in an old stub or stump not six feet high, and which we have seen and passed several times without giving it a thought. We go farther down the mountain and beat about to the right and left, and get entangled in brush and arrested by precipices, and finally, as the day is nearly spent, give up the search and leave the woods quite baffled, but resolved to return on the morrow. The next day we come back and commence operations in an opening in the woods well down on the side of the mountain where we gave up the search. Our box is soon swarming with the eager bees, and they go back toward the summit we have passed. We follow back and establish a new line, where the ground will permit; then another and still another, and yet the riddle is not solved. One time we are south of them, then north, then the bees get up through the trees and we cannot tell where they go. But after much searching, and after the mystery seems rather to deepen than to clear up, we chance to pause beside the old stump. A bee comes out of a small opening like that made by ants in decayed wood, rubs its eyes and examines its antennæ, as bees always do before leaving their hive, then takes flight. At the same instant several bees come by us loaded with our honey and settle home with that peculiar low, complacent buzz of the well-filled insect. Here, then, is our idyl, our bit of Virgil and Theocritus, in a decayed stump of a hemlock-tree. We could tear it open with our hands, and a bear would find it an easy prize, and a rich one, too, for we take from it fifty pounds of excellent honey. The bees have been here many years, and have of course sent out swarm after swarm into the wilds. They have protected themselves against the weather and strengthened their shaky habitation by a copious use of wax.

When a bee-tree is thus "taken up" in the middle of the day, of course a good many bees are away from home and have not heard the news. When they return and find the ground flowing with honey, and piles of bleeding combs lying about, they appar-

ently do not recognize the place, and their first
instinct is to fall to and fill themselves; this done,
their next thought is to carry it home, so they rise
up slowly through the branches of the trees till
they have attained an altitude that enables them to
survey the scene, when they seem to say, "Why,
this is home," and down they come again; beholding the wreck and ruins once more, they still think
there is some mistake, and get up a second or a third
time and then drop back pitifully as before. It is
the most pathetic sight of all, the surviving and
bewildered bees struggling to save a few drops of
their wasted treasures.

Presently, if there is another swarm in the woods,
robber bees appear. You may know them by their
saucy, chiding, devil-may-care hum. It is an ill
wind that blows nobody good, and they make the
most of the misfortune of their neighbors, and
thereby pave the way for their own ruin. The
hunter marks their course, and the next day looks
them up. On this occasion the day was hot and
the honey very fragrant, and a line of bees was soon
established south-southwest. Though there was
much refuse honey in the old stub, and though
little golden rills trickled down the hill from it, and
the near branches and saplings were besmeared with
it where we wiped our murderous hands, yet not a
drop was wasted. It was a feast to which not only
honey-bees came, but bumblebees, wasps, hornets

flies, ants. The bumblebees, which at this season are hungry vagrants with no fixed place of abode, would gorge themselves, then creep beneath the bits of empty comb or fragments of bark and pass the night, and renew the feast next day. The bumble-bee is an insect of which the bee-hunter sees much. There are all sorts and sizes of them. They are dull and clumsy compared with the honey-bee. Attracted in the fields by the bee-hunter's box, they will come up the wind on the scent and blunder into it in the most stupid, lubberly fashion.

The honey-bees that licked up our leavings on the old stub belonged to a swarm, as it proved, about half a mile farther down the ridge, and a few days afterward fate overtook them, and their stores in turn became the prey of another swarm in the vicinity, which also tempted Providence and were overwhelmed. The first-mentioned swarm I had lined from several points, and was following up the clew over rocks and through gullies, when I came to where a large hemlock had been felled a few years before, and a swarm taken from a cavity near the top of it; fragments of the old comb were yet to be seen. A few yards away stood another short, squatty hemlock, and I said my bees ought to be there. As I paused near it, I noticed where the tree had been wounded with an axe a couple of feet from the ground many years before. The wound had partially grown over, but there was an opening there that I did not see at the first glance. I was about to pass on when a bee passed me making that peculiar shrill, discordant hum that a bee makes when besmeared with honey. I saw it alight in the partially closed wound and crawl home ; then came others and others, little bands and squads of them, heavily freighted with honey from the box. The tree was about twenty inches through and hollow at the butt, or from the axe-mark down. This space the bees had completely filled with honey. With an axe we cut away the outer ring of live wood and exposed the treasure. Despite the utmost care, we wounded the comb so that little rills of the golden liquid issued from the root of the tree and trickled down the hill.

The other bee-tree in the vicinity to which I have referred we found one warm November day in less than half an hour after entering the woods. It also was a hemlock, that stood in a niche in a wall of hoary moss-covered rocks thirty feet high. The tree hardly reached to the top of the precipice. The bees entered a small hole at the root, which was seven or eight feet from the ground. The position was a striking one. Never did apiary have a finer outlook or more rugged surroundings. A black, wood-embraced lake lay at our feet; the long panorama of the Catskills filled the far distance, and the more broken outlines of the Shawangunk range filled the rear. On every hand were precipices and a wild confusion of rocks and trees.

The cavity occupied by the bees was about three feet and a half long and eight or ten inches in diameter. With an axe we cut away one side of the tree, and laid bare its curiously wrought heart of honey. It was a most pleasing sight. What winding and devious ways the bees had through their palace! What great masses and blocks of snow-white comb there were! Where it was sealed up, presenting that slightly dented, uneven surface, it looked like some precious ore. When we carried a large pailful of it out of the woods. it seemed still more like ore.

Your native bee-hunter predicates the distance of the tree by the time the bee occupies in making its first trip. But this is no certain guide. You are always safe in calculating that the tree is inside of a mile, and you need not as a rule look for your bee's return under ten minutes. One day I picked up a bee in an opening in the woods and gave it honey, and it made three trips to my box with an interval of about twelve minutes between them; it returned alone each time ; the tree, which I afterward found, was about half a mile distant.

In lining bees through the woods, the tactics of the hunter are to pause every twenty or thirty rods, lop away the branches or cut down the trees, and set the bees to work again. If they still go forward, he goes forward also, and repeats his observations till the tree is found, or till the bees turn and come back upon the trail. Then he knows he has passed the tree, and he retraces his steps to a convenient distance and tries again, and thus quickly reduces the space to be looked over till the swarm is traced home. On one occasion, in a wild rocky wood, where the surface alternated between deep gulfs and chasms filled with thick, heavy growths of timber, and sharp, precipitous, rocky ridges like a tempest-tossed sea, I carried my bees directly under their tree, and set them to work from a high, exposed ledge of rocks not thirty feet distant. One would have expected them under such circumstances to have gone straight home, as there were but few branches intervening, but they did not; they labored up through the trees and attained an altitude above the woods as if they had miles to travel, and thus baffled me for hours. Bees will always do this. They are acquainted with the woods only from the top side, and from the air above; they recognize home only by landmarks here, and in every instance they rise aloft to take their bearings. Think how familiar to them the topography of the forest summits must be, — an umbrageous sea or plain where every mark and point is known.

Another curious fact is that generally you will get track of a bee-tree sooner when you are half a mile from it than when you are only a few yards.

Bees, like us human insects, have little faith in the near at hand; they expect to make their fortune in a distant field, they are lured by the remote and the difficult, and hence overlook the flower and the sweet at their very door. On several occasions I have unwittingly set my box within a few paces of a bee-tree and waited long for bees without getting them, when, on removing to a distant field or opening in the woods, I have got a clew at once.

I have a theory that when bees leave the hive, unless there is some special attraction in some other direction, they generally go against the wind. They would thus have the wind with them when they returned home heavily laden, and with these little navigators the difference is an important one. With a full cargo, a stiff head-wind is a great hindrance, but fresh and empty-handed, they can face it with more ease. Virgil says bees bear gravel-stones as ballast, but their only ballast is their honey-bag. Hence, when I go bee-hunting, I prefer to get to windward of the woods in which the swarm is supposed to have refuge.

Bees, like the milkman, like to be near a spring. They do water their honey, especially in a dry time. The liquid is then of course thicker and sweeter, and will bear diluting. Hence old bee-hunters look for bee-trees along creeks and near spring runs in the woods. I once found a tree a long distance from any water, and the honey had a peculiar bitter flavor, imparted to it, I was convinced, by rainwater sucked from the decayed and spongy hemlock-tree in which the swarm was found. In cutting into the tree, the north side of it was found to be saturated with water like a spring, which ran out in big drops, and had a bitter flavor. The bees had thus found a spring or a cistern in their own house.

Bees are exposed to many hardships and many dangers. Winds and storms prove as disastrous to them as to other navigators. Black spiders lie in wait for them as do brigands for travelers. One day, as I was looking for a bee amid some goldenrod, I spied one partly concealed under a leaf. Its baskets were full of pollen, and it did not move. On lifting up the leaf I discovered that a hairy spider was ambushed there and had the bee by the throat. The vampire was evidently afraid of the bee's sting, and was holding it by the throat till quite sure of its death. Virgil speaks of the painted lizard, perhaps a species of salamander, as an enemy of the honey-bee. We have no lizard that destroys the bee; but our tree-toad, ambushed among the apple and cherry blossoms, snaps them up wholesale. Quick as lightning that subtle but clammy tongue darts forth, and the unsuspecting bee is gone. Virgil also accuses the titmouse and the woodpecker of preying upon the bees, and our kingbird has been charged with the like crime, but the latter devours only the drones. The workers are either too small and quick for it or else it dreads their sting.

Virgil, by the way, had little more than a child's knowledge of the honey-bee. There is little fact and much fable in his fourth Georgic. If he had ever kept bees himself, or even visited an apiary, it is hard to see how he could have believed that the bee in its flight abroad carried a gravel-stone for ballast:

"And as when empty barks on billows float,
  With sandy ballast sailors trim the boat;
  So bees bear gravel-stones, whose poising weight
  Steers through the whistling winds their steady flight;"

or that, when two colonies made war upon each other, they issued forth from their hives led by their kings and fought in the air, strewing the ground with the dead and dying: —

"Hard hailstones lie not thicker on the plain,
  Nor shaken oaks such show'rs of acorns rain."

It is quite certain he had never been bee-hunting. If he had, we should have had a fifth Georgic. Yet he seems to have known that bees sometimes escaped to the woods: —

"Nor bees are lodged in hives alone, but found
  In chambers of their own beneath the ground:
  Their vaulted roofs are hung in pumices,
  And in the rotten trunks of hollow trees."

AN IDYL OF THE HONEY-BEE

Wild honey is as near like tame as wild bees are like their brothers in the hive. The only difference is, that wild honey is flavored with your adventure, which makes it a little more delectable than the domestic article.

A BUNCH OF HERBS

Spotting a wildflower in front of Slabsides

EDITOR'S NOTE

"A Bunch of Herbs" conveys a sense of what it might be like to go on a nature walk with Burroughs. Many who did walk with him regularly reported that they always learned something new or saw something familiar in a new way. In this essay, Burroughs's attentive descriptions of ordinary wildflowers and weeds, like the moth mullein or dandelion, remind the reader that nothing is insignificant simply because it is small or common or because its human usefulness is unknown. His knowledge of ecological change also reminds the reader how little of a so-called native landscape is made up of indigenous plants. "We have hardly a weed we can call our own," Burroughs writes; our weeds and flowers, like most Americans, are descendants of foreign invaders. Burroughs's imaginative reading of an ordinary landscape tells a dramatic story of immigration and change. His love of natural smells, evident in his comparison of English and American wildflowers, extended to the evanescent and even to what others might consider rank. His biographer, Clara Barrus, reports that Burroughs seldom walked

EDITOR'S NOTE

past a clump of tansy without plucking a leaf from the weed, "crushing it the better to smell it, then poking it in one of his nostrils, whence it comically protruded as he continued his walk." This essay appeared in *Pepacton* in 1881 and was reprinted in *A Bunch of Herbs and Other Papers* as part of the Riverside Literature Series in 1896.

F.B.

A BUNCH OF HERBS

FRAGRANT WILD FLOWERS

THE charge that was long ago made against our wild flowers by English travelers in this country, namely, that they are odorless, doubtless had its origin in the fact that, whereas in England the sweet-scented flowers are among the most common and conspicuous, in this country they are rather shy and withdrawn, and consequently not such as travelers would be likely to encounter. Moreover, the British traveler, remembering the deliciously fragrant blue violets he left at home, covering every grassy slope and meadow bank in spring, and the wild clematis. or traveler's joy, overrunning hedges and old walls with its white, sweet-scented blossoms, and finding the corresponding species here equally abundant but entirely scentless, very naturally infers that our wild flowers are all deficient in this respect. He would be confirmed in this opinion when, on turning to some of our most beautiful and striking native flowers, like the laurel, the rhododendron, the columbine, the inimitable fringed gentian, the burning cardinal-flower, or our asters and goldenrod, dashing the roadsides with tints of purple and gold, he found them scentless also. "Where are your fragrant flowers?" he might well say; "I can find none." Let him look closer and penetrate our forests, and visit our ponds and lakes. Let him compare our matchless, rosy-lipped, honey-hearted trailing arbutus with his own ugly ground-ivy; let him compare our sumptuous, fragrant pond-lily with his own odorless *Nymphœa alba*. In our Northern woods he will find the floors carpeted with the delicate linnæa, its twin rose-colored, nodding flowers filling the air with fragrance. (I am aware that the linnæa is found in some parts of Northern Europe.) The fact is, we perhaps have as many sweet-scented wild flowers as Europe has, only they are not quite so prominent in our flora, nor so well known to our people or to our poets.

Think of Wordsworth's "Golden Daffodils:" —

"I wandered lonely as a cloud
    That floats on high o'er vales and hills,
When, all at once, I saw a crowd,
    A host of golden daffodils,
Beside the lake, beneath the trees,
Fluttering and dancing in the breeze.

"Continuous as the stars that shine
    And twinkle on the milky way,

A BUNCH OF HERBS

> They stretched in never-ending line
> Along the margin of a bay.
> Ten thousand saw I at a glance,
> Tossing their heads in sprightly dance."

No such sight could greet the poet's eye here. He might see ten thousand marsh marigolds, or ten times ten thousand houstonias, but they would not toss in the breeze, and they would not be sweet-scented like the daffodils.

It is to be remembered, too, that in the moister atmosphere of England the same amount of fragrance would be much more noticeable than with us. Think how our sweet bay, or our pink azalea, or our white alder, to which they have nothing that corresponds, would perfume that heavy, vapor-laden air!

In the woods and groves in England, the wild hyacinth grows very abundantly in spring, and in places the air is loaded with its fragrance. In our woods a species of dicentra, commonly called squirrel corn, has nearly the same perfume, and its racemes of nodding whitish flowers, tinged with pink, are quite as pleasing to the eye, but it is a shyer, less abundant plant. When our children go to the fields in April and May, they can bring home no wild flowers as pleasing as the sweet English violet, and cowslip, and yellow daffodil, and wallflower; and when British children go to the woods at the same season, they can load their hands and baskets with nothing that compares with our trailing arbutus, or, later in the season, with our azaleas; and, when their boys go fishing or boating in summer, they can wreathe themselves with nothing that approaches our pond-lily.

There are upward of thirty species of fragrant native wild flowers and flowering shrubs and trees in New England and New York, and, no doubt, many more in the South and West. My list is as follows: —

White violet (*Viola blanda*).
Canada violet (*Viola Canadensis*).
Hepatica (occasionally fragrant).
Trailing arbutus (*Epigæa repens*).
Mandrake (*Podophyllum peltatum*).
Yellow lady's-slipper (*Cypripedium parviflorum*).
Purple lady's-slipper (*Cypripedium acaule*).
Squirrel corn (*Dicentra Canadensis*).
Showy orchis (*Orchis spectabilis*).
Purple fringed-orchis (*Habenaria psycodes*).
Arethusa (*Arethusa bulbosa*).
Calopogon (*Calopogon pulchellus*).
Lady's-tresses (*Spiranthes cernua*).
Pond-lily (*Nymphæa odorata*).
Wild rose (*Rosa nitida*).
Twin-flower (*Linnæa borealis*).
Sugar maple (*Acer saccharinum*).
Linden (*Tilia Americana*).

A BUNCH OF HERBS

Locust-tree (*Robinia pseudacacia*).
White alder (*Clethra alnifolia*).
Smooth azalea (*Rhododendron arborescens*).
White azalea (*Rhododendron viscosum*).
Pinxter-flower (*Rhododendron nudiflorum*).
Yellow azalea (*Rhododendron calendulaceum*).
Sweet bay (*Magnolia glauca*).
Mitchella vine (*Mitchella repens*).
Sweet coltsfoot (*Petasites palmata*).
Pasture thistle (*Cnicus pumilus*).
False wintergreen (*Pyrola rotundifolia*).
Spotted wintergreen (*Chimaphila maculata*).
Prince's pine (*Chimaphila umbellata*).
Evening primrose (*Œnothera biennis*).
Hairy loosestrife (*Steironema ciliatum*).
Dogbane (*Apocynum*).
Ground-nut (*Apios tuberosa*).
Adder's-tongue pogonia (*Pogonia ophioglossoides*).
Wild grape (*Vitis cordofolia*).
Horned bladderwort (*Utricularia cornuta*).

The last-named, horned bladderwort, is perhaps the most fragrant flower we have. In a warm, moist atmosphere, its odor is almost too strong. It is a plant with a slender, leafless stalk or scape less than a foot high, with two or more large yellow hood or helmet shaped flowers. It is not common, and belongs pretty well north, growing in sandy swamps and along the marshy margins of lakes and ponds. Its perfume is sweet and spicy in an emi-

nent degree. I have placed in the above list several
flowers that are intermittently fragrant, like the
hepatica, or liver-leaf. This flower is the earliest,
as it is certainly one of the most beautiful, to be
found in our woods, and occasionally it is fragrant.
Group after group may be inspected, ranging
through all shades of purple and blue, with some
perfectly white, and no odor be detected, when presently you will happen upon a little brood of them
that have a most delicate and delicious fragrance.
The same is true of a species of loosestrife growing
along streams and on other wet places, with tall
bushy stalks, dark green leaves, and pale axillary
yellow flowers (probably European). A handful of
these flowers will sometimes exhale a sweet fragrance; at other times, or from another locality,
they are scentless. Our evening primrose is thought
to be uniformly sweet-scented, but the past season
I examined many specimens, and failed to find one
that was so. Some seasons the sugar maple yields
much sweeter sap than in others; and even individual trees, owing to the soil, moisture, and other
conditions where they stand, show a great difference in this respect. The same is doubtless true of
the sweet-scented flowers. I had always supposed
that our Canada violet — the tall, leafy-stemmed
white violet of our Northern woods — was odorless,
till a correspondent called my attention to the contrary fact. On examination I found that, while the

first ones that bloomed about May 25 had very sweet-scented foliage, especially when crushed in the hand, the flowers were practically without fragrance. But as the season advanced the fragrance developed, till a single flower had a well-marked perfume, and a handful of them was sweet indeed. A single specimen, plucked about August 1, was quite as fragrant as the English violet, though the perfume is not what is known as violet, but, like that of the hepatica, comes nearer to the odor of certain fruit trees.

It is only for a brief period that the blossoms of our sugar maple are sweet-scented; the perfume seems to become stale after a few days: but pass under this tree just at the right moment, say at nightfall on the first or second day of its perfect inflorescence, and the air is laden with its sweetness; its perfumed breath falls upon you as its cool shadow does a few weeks later.

After the linnæa and the arbutus, the prettiest sweet-scented flowering vine our woods hold is the common mitchella vine, called squaw-berry and partridge-berry. It blooms in June, and its twin flowers, light cream-color, velvety, tubular, exhale a most agreeable fragrance.

Our flora is much more rich in orchics than the European, and many of ours are fragrant. The first to bloom in the spring is the showy orchis, though it is far less showy than several others. I find it in May, not on hills, where Gray says it grows, but in low, damp places in the woods. It has two oblong shining leaves, with a scape four or five inches high strung with sweet-scented, pink-purple flowers. I usually find it and the fringed polygala in bloom at the same time; the lady's-slipper is a little later. The purple fringed-orchis, one of the most showy and striking of all our orchids, blooms in midsummer in swampy meadows and in marshy, grassy openings in the woods, shooting up a tapering column or cylinder of pink-purple fringed flowers, that one may see at quite a distance, and the perfume of which is too rank for a close room. This flower is, perhaps, like the English fragrant orchis, found in pastures.

Few fragrant flowers in the shape of weeds have come to us from the Old World, and this leads me to remark that plants with sweet-scented flowers are, for the most part, more intensely local, more fastidious and idiosyncratic, than those without perfume. Our native thistle — the pasture thistle — has a marked fragrance, and it is much more shy and limited in its range than the common Old World thistle that grows everywhere. Our little, sweet white violet grows only in wet places, and the Canada violet only in high, cool woods, while the common blue violet is much more general in its distribution. How fastidious and exclusive is the cypripedium! You will find it in one locality in the woods, usually on high, dry ground, and will look in vain for it elsewhere. It does not go in herds like the commoner plants, but affects privacy and solitude. When I come upon it in my walks, I seem to be intruding upon some very private and exclusive company. The large yellow cypripedium has a peculiar, heavy, oily odor.

In like manner one learns where to look for arbutus, for pipsissewa, for the early orchis; they have their particular haunts, and their surroundings are nearly always the same. The yellow pond-lily is found in every sluggish stream and pond, but *Nymphœa odorata* requires a nicer adjustment of conditions, and consequently is more restricted in its range. If the mullein were fragrant, or toad-flax, or the daisy, or blue-weed, or goldenrod, they would doubtless be far less troublesome to the agriculturist. There are, of course, exceptions to the rule I have here indicated, but it holds in most cases. Genius is a specialty: it does not grow in every soil; it skips the many and touches the few; and the gift of perfume to a flower is a special grace like genius or like beauty, and never becomes common or cheap.

"Do honey and fragrance always go together in the flowers?" Not uniformly. Of the list of fragrant wild flowers I have given, the only ones that the bees procure nectar from, so far as I have observed, are arbutus, dicentra, sugar maple, locust,

and linden. Non-fragrant flowers that yield honey
are those of the raspberry, clematis, sumac, white
oak, bugloss, ailanthus, goldenrod, aster, fleabane.
A large number of odorless plants yield pollen to
the bee. There is nectar in the columbine, and the
bumblebee sometimes gets it by piercing the spur
from the outside as she does with dicentra. There
ought to be honey in the honeysuckle, but I have
never seen the hive-bee make any attempt to get it.

WEEDS

One is tempted to say that the most human
plants, after all, are the weeds. How they cling
to man and follow him around the world, and
spring up wherever he sets his foot! How they
crowd around his barns and dwellings, and throng
his garden and jostle and override each other in
their strife to be near him! Some of them are so
domestic and familiar, and so harmless withal, that
one comes to regard them with positive affection.
Motherwort, catnip, plantain, tansy, wild mustard,
— what a homely human look they have! they are
an integral part of every old homestead. Your smart
new place will wait long before they draw near it
Or knot-grass, that carpets every old dooryard, and
fringes every walk, and softens every path that
knows the feet of children, or that leads to the
spring, or to the garden, or to the barn, how kindly
one comes to look upon it! Examine it with a

pocket glass and see how wonderfully beautiful and exquisite are its tiny blossoms. It loves the human foot, and when the path or the place is long disused, other plants usurp the ground.

The gardener and the farmer are ostensibly the greatest enemies of the weeds, but they are in reality their best friends. Weeds, like rats and mice, increase and spread enormously in a cultivated country. They have better food, more sunshine, and more aids in getting themselves disseminated. They are sent from one end of the land to the other in seed grain of various kinds, and they take their share, and more too, if they can get it, of the phosphates and stable manures. How sure, also, they are to survive any war of extermination that is waged against them! In yonder field are ten thousand and one Canada thistles. The farmer goes resolutely to work and destroys ten thousand and thinks the work is finished, but he has done nothing till he has destroyed the ten thousand and one. This one will keep up the stock and again cover his field with thistles.

Weeds are Nature's makeshift. She rejoices in the grass and the grain, but when these fail to cover her nakedness she resorts to weeds. It is in her plan or a part of her economy to keep the ground constantly covered with vegetation of some sort, and she has layer upon layer of seeds in the soil for this purpose, and the wonder is that each kind lies dormant until it is wanted. If I uncover the earth in any of my fields, ragweed and pigweed spring up; if these are destroyed, harvest grass, or quack grass, or purslane, appears. The spade or the plow that turns these under is sure to turn up some other variety, as chickweed, sheep-sorrel, or goose-foot. The soil is a storehouse of seeds.

The old farmers say that wood-ashes will bring in the white clover, and they will; the germs are in the soil wrapped in a profound slumber, but this stimulus tickles them until they awake. Stramonium has been known to start up on the site of an old farm building, when it had not been seen in that locality for thirty years. I have been told that a farmer, somewhere in New England, in digging a well came at a great depth upon sand like that of the seashore; it was thrown out, and in due time there sprang from it a marine plant. I have never seen earth taken from so great a depth that it would not before the end of the season be clothed with a crop of weeds. Weeds are so full of expedients, and the one engrossing purpose with them is to multiply. The wild onion multiplies at both ends, — at the top by seed, and at the bottom by offshoots. Toad-flax travels under ground and above ground. Never allow a seed to ripen, and yet it will cover your field. Cut off the head of the wild carrot, and in a week or two there are five heads in place of this one; cut off these, and by fall there are ten looking defiance at you from the same root. Plant corn in August, and it will go forward with its preparations as if it had the whole season before it. Not so with the weeds; they have learned better. If amaranth, or abutilon, or burdock gets a late start, it makes great haste to develop its seed; it foregoes its tall stalk and wide flaunting growth, and turns all its energies into keeping up the succession of the species. Certain fields under the plow are always infested with "blind nettles," others with wild buckwheat, black bindweed, or cockle. The seed lies dormant under the sward, the warmth and the moisture affect it not until other conditions are fulfilled.

The way in which one plant thus keeps another down is a great mystery. Germs lie there in the soil and resist the stimulating effect of the sun and the rains for years, and show no sign. Presently something whispers to them, "Arise, your chance has come; the coast is clear;" and they are up and doing in a twinkling.

Weeds are great travelers; they are, indeed, the tramps of the vegetable world. They are going east, west, north, south; they walk; they fly; they swim ; they steal a ride ; they travel by rail, by flood, by wind ; they go under ground, and they go above, across lots, and by the highway. But, like other tramps, they find it safest by the highway: in the fields they are intercepted and cut off;

but on the public road, every boy, every passing herd of sheep or cows, gives them a lift. Hence the incursion of a new weed is generally first noticed along the highway or the railroad. In Orange County I saw from the car window a field overrun with what I took to be the branching white mullein. Gray says it is found in Pennsylvania and at the head of Oneida Lake. Doubtless it had come by rail from one place or the other. Our botanist says of the bladder campion, a species of pink, that it has been naturalized around Boston ; but it is now much farther west, and I know fields along the Hudson overrun with it. Streams and water-courses are the natural highway of the weeds. Some years ago, and by some means or other, the viper's bugloss, or blue-weed, which is said to be a troublesome weed in Virginia, effected a lodgment near the head of the Esopus Creek, a tributary of the Hudson. From this point it has made its way down the stream, overrunning its banks and invading meadows and cultivated fields, and proving a serious obstacle to the farmer. All the gravelly, sandy margins and islands of the Esopus, sometimes acres in extent, are in June and July blue with it, and rye and oats and grass in the near fields find it a serious competitor for possession of the soil. It has gone down the Hudson, and is appearing in the fields along its shores. The tides carry it up the mouths of the streams where it takes root; the

winds, or the birds, or other agencies, in time give
it another lift, so that it is slowly but surely making
its way inland. The bugloss belongs to what may
be called beautiful weeds, despite its rough and
bristly stalk. Its flowers are deep violet-blue, the
stamens exserted, as the botanists say, that is, projected beyond the mouth of the corolla, with showy
red anthers. This bit of red, mingling with the
blue of the corolla, gives a very rich, warm purple
hue to the flower, that is especially pleasing at a
little distance. The best thing I know about this
weed besides its good looks is that it yields honey
or pollen to the bee.

Another foreign plant that the Esopus Creek has
distributed along its shores and carried to the Hudson is saponaria, known as "Bouncing Bet." It is
a common and in places a troublesome weed in this
valley. Bouncing Bet is, perhaps, its English name,
as the pink-white complexion of its flowers with
their perfume and the coarse, robust character of
the plant really give it a kind of English feminine
comeliness and bounce. It looks like a Yorkshire
housemaid. Still another plant in my section, which
I notice has been widely distributed by the agency
of water, is the spiked loosestrife. It first appeared
many years ago along the Wallkill ; now it may
be seen upon many of its tributaries and all along
its banks ; and in many of the marshy bays and
coves along the Hudson, its great masses of pur-

ple-red bloom in middle and late summer affording
a welcome relief to the traveler's eye. It also belongs to the class of beautiful weeds. It grows rank
and tall, in dense communities, and always presents to the eye a generous mass of color. In places,
the marshes and creek banks are all aglow with it, its
wand-like spikes of flowers shooting up and uniting
in volumes or pyramids of still flame. Its petals,
when examined closely, present a curious wrinkled
or crumpled appearance, like newly washed linen;
but when massed, the effect is eminently pleasing.
It also came from abroad, probably first brought
to this country as a garden or ornamental plant.

As a curious illustration of how weeds are carried
from one end of the earth to the other, Sir Joseph
Hooker relates this circumstance : " On one occasion," he says, "landing on a small uninhabited
island nearly at the Antipodes, the first evidence
I met with of its having been previously visited by
man was the English chickweed ; and this I traced
to a mound that marked the grave of a British
sailor, and that was covered with the plant, doubtless the offspring of seed that had adhered to the
spade or mattock with which the grave had been
dug."

Ours is a weedy country because it is a roomy
country. Weeds love a wide margin, and they find
it here. You shall see more weeds in one day's
travel in this country than in a week's journey in

A BUNCH OF HERBS

Europe. Our culture of the soil is not so close and thorough, our occupancy not so entire and exclusive. The weeds take up with the farmers' leavings, and find good fare. One may see a large slice taken from a field by elecampane, or by teasel or milkweed; whole acres given up to whiteweed, goldenrod, wild carrots, or the ox-eye daisy; meadows overrun with bear-weed, and sheep pastures nearly ruined by St. John's-wort or the Canada thistle. Our farms are so large and our husbandry so loose that we do not mind these things. By and by we shall clean them out. When Sir Joseph Hooker landed in New England a few years ago, he was surprised to find how the European plants flourished there. He found the wild chicory growing far more luxuriantly than he had ever seen it elsewhere, "forming a tangled mass of stems and branches, studded with turquoise-blue blossoms, and covering acres of ground." This is one of the many weeds that Emerson binds into a bouquet in his "Humble-Bee:" —

> "Succory to match the sky,
> Columbine with horn of honey,
> Scented fern and agrimony,
> Clover, catchfly, adder's-tongue,
> And brier-roses, dwelt among."

A less accurate poet than Emerson would probably have let his reader infer that the bumblebee gathered

honey from all these plants, but Emerson is careful
to say only that she dwelt among them. Succory is
one of Virgil's weeds also, —

"And spreading succ'ry chokes the rising field."

Is there not something in our soil and climate
exceptionally favorable to weeds, — something
harsh, ungenial, sharp-toothed, that is akin to them?
How woody and rank and fibrous many varieties
become, lasting the whole season, and standing up
stark and stiff through the deep winter snows, —
desiccated, preserved by our dry air! Do nettles
and thistles bite so sharply in any other country?
Let the farmer tell you how they bite of a dry midsummer day when he encounters them in his wheat
or oat harvest.

Yet it is a fact that all our more pernicious weeds,
like our vermin, are of Old World origin. They hold
up their heads and assert themselves here, and take
their fill of riot and license; they are avenged for
their long years of repression by the stern hand of
European agriculture. We have hardly a weed we
can call our own. I recall but three that are at all
noxious or troublesome, namely, milkweed, ragweed, and goldenrod; but who would miss the last
from our fields and highways?

"Along the roadside, like the flowers of gold
    That tawny Incas for their gardens wrought,
Heavy with sunshine droops the goldenrod,"

sings Whittier. In Europe our goldenrod is cultivated in the flower gardens, as well it may be. The
native species is found mainly in woods, and is much
less showy than ours.

Our milkweed is tenacious of life; its roots lie
deep, as if to get away from the plow, but it seldom
infests cultivated crops. Then its stalk is so full
of milk and its pod so full of silk that one cannot
but ascribe good intentions to it, if it does sometimes overrun the meadow.

> "In dusty pods the milkweed
>   Its hidden silk has spun,"

sings "H. H." in her "September."

Of our ragweed not much can be set down that
is complimentary, except that its name in the botany is *Ambrosia*, food of the gods. It must be the
food of the gods if anything, for, so far as I have
observed, nothing terrestrial eats it, not even billy-
goats. (Yet a correspondent writes me that in
Kentucky the cattle eat it when hard-pressed, and
that a certain old farmer there, one season when
the hay crop failed, cut and harvested tons of it for
his stock in winter. It is said that the milk and
butter made from such hay are not at all suggestive
of the traditional Ambrosia!) It is the bane of
asthmatic patients, but the gardener makes short
work of it. It is about the only one of our weeds that
follows the plow and the harrow, and, except that

it is easily destroyed, I should suspect it to be an immigrant from the Old World. Our fleabane is a troublesome weed at times, but good husbandry has little to dread from it.

But all the other outlaws of the farm and garden come to us from over seas; and what a long list it is:—

Common thistle,	Gill,
Canada thistle,	Nightshade,
Burdock,	Buttercup,
Yellow dock,	Dandelion,
Wild carrot,	Wild mustard,
Ox-eye daisy,	Shepherd's purs
Chamomile,	St. John's-wort
Mullein,	Chickweed,
Dead-nettle (*Lamium*),	Purslane,
Hemp nettle (*Galeopsis*),	Mallow,
Elecampane,	Darnel,
Plantain,	Poison hemlock,
Motherwort,	Hop-clover,
Stramonium,	Yarrow,
Catnip,	Wild radish,
Blue-weed,	Wild parsnip,
Stick-seed,	Chicory,
Hound's-tongue,	Live-forever,
Henbane,	Toad-flax
Pigweed,	Sheep-sorrel,
Quitch grass,	Mayweed,

and others less noxious. To offset this list we have

given Europe the vilest of all weeds, a parasite that sucks up human blood, tobacco. Now if they catch the Colorado beetle of us, it will go far toward paying them off for the rats and the mice, and for other pests in our houses.

The more attractive and pretty of the British weeds — as the common daisy, of which the poets have made so much, the larkspur, which is a pretty cornfield weed, and the scarlet field-poppy, which flowers all summer, and is so taking amid the ripening grain — have not immigrated to our shores. Like a certain sweet rusticity and charm of European rural life, they do not thrive readily under our skies. Our fleabane has become a common roadside weed in England, and a few other of our native less known plants have gained a foothold in the Old World. Our beautiful jewel-weed has recently appeared along certain of the English rivers.

Pokeweed is a native American, and what a lusty, royal plant it is! It never invades cultivated fields, but hovers about the borders and looks over the fences like a painted Indian sachem. Thoreau coveted its strong purple stalk for a cane, and the robins eat its dark crimson-juiced berries.

It is commonly believed that the mullein is indigenous to this country, for have we not heard that it is cultivated in European gardens, and christened the American velvet plant? Yet it, too, seems to have come over with the Pilgrims, and is most abundant in the older parts of the country. It abounds throughout Europe and Asia, and had its economic uses with the ancients. The Greeks made lamp-wicks of its dried leaves, and the Romans dipped its dried stalk in tallow for funeral torches. It affects dry uplands in this country, and, as it takes two years to mature, it is not a troublesome weed in cultivated crops. The first year it sits low upon the ground in its coarse flannel leaves, and makes ready ; if the plow comes along now, its career is ended. The second season it starts upward its tall stalk, which in late summer is thickly set with small yellow flowers, and in fall is charged with myriads of fine black seeds. "As full as a dry mullein stalk of seeds" is almost equivalent to saying, "as numerous as the sands upon the seashore."

Perhaps the most notable thing about the weeds that have come to us from the Old World, when compared with our native species, is their persistence, not to say pugnacity. They fight for the soil; they plant colonies here and there, and will not be rooted out. Our native weeds are for the most part shy and harmless, and retreat before cultivation, but the European outlaws follow man like vermin; they hang to his coat-skirts, his sheep transport them in their wool, his cow and horse in tail and mane. As I have before said, it is as with the rats and mice. The American rat is in the woods and is rarely seen even by woodmen, and the native mouse barely hovers upon the outskirts of civilization; while the Old World species defy our traps and our poison, and have usurped the land. So with the weeds. Take the thistle for instance: the common and abundant one everywhere, in fields and along highways, is the European species; while the native thistles, swamp thistle, pasture thistle, etc., are much more shy, and are not at all troublesome. The Canada thistle, too, which came to us by way of Canada, — what a pest, what a usurper, what a defier of the plow and the harrow! I know of but one effectual way to treat it, — put on a pair of buckskin gloves, and pull up every plant that shows itself; this will effect a radical cure in two summers. Of course the plow or the scythe, if not allowed to rest more than a month at a time, will finally conquer it.

Or take the common St. John's-wort, — how it has established itself in our fields and become a most pernicious weed, very difficult to extirpate ; while the native species are quite rare, and seldom or never invade cultivated fields, being found mostly in wet and rocky waste places. Of Old World origin, too, is the curled-leaf dock that is so annoying about one's garden and home meadows, its long tapering root clinging to the soil with such tenacity that I have pulled upon it till I could see stars without budging it; it has more lives than a cat, making a shift to live when pulled up and laid on top of the ground in the burning summer sun. Our native docks are mostly found in swamps, or near them, and are harmless.

Purslane — commonly called "pusley," and which has given rise to the saying, "as mean as pusley" — of course is not American. A good sample of our native purslane is the claytonia, or spring beauty, a shy, delicate plant that opens its rose-colored flowers in the moist, sunny places in the woods or along their borders so early in the season.

There are few more obnoxious weeds in cultivated ground than sheep-sorrel, also an Old World plant; while our native wood-sorrel, with its white, delicately veined flowers, or the variety with yellow flowers, is quite harmless. The same is true of the mallow, the vetch, the tare, and other plants. We have no native plant so indestructible as garden orpine, or live-forever, which our grandmothers nursed, and for which they are cursed by many a farmer. The fat, tender, succulent dooryard stripling turned out to be a monster that would devour the earth. I have seen acres of meadow land destroyed by it. The way to drown an amphibious animal is never to allow it to come to the surface to breathe, and this is the way to kill live-forever. It lives by its stalk and leaf, more than by its root, and, if cropped or bruised as soon as it comes to the surface, it will in time perish. It laughs the plow the hoe, the cultivator to scorn, but grazing herds will eventually scotch it. Our two species of native orpine, *Sedum ternatum* and *S. telephioides*, are never troublesome as weeds.

The European weeds are sophisticated, domesticated, civilized ; they have been to school to man for many hundred years, and they have learned to thrive upon him : their struggle for existence has been sharp and protracted; it has made them hardy and prolific ; they will thrive in a lean soil, or they will wax strong in a rich one; in all cases they follow man and profit by him. Our native weeds, on the other hand, are furtive and retiring; they flee before the plow and the scythe, and hide in corners and remote waste places. Will they, too, in time, change their habits in this respect?

"Idle weeds are fast in growth," says Shakespeare, but that depends upon whether the competition is sharp and close. If the weed finds itself distanced, or pitted against great odds, it grows more slowly and is of diminished stature, but let it once get the upper hand, and what strides it makes! Red-root will grow four or five feet high if it has a chance, or it will content itself with a few inches and mature its seed almost upon the ground.

Many of our worst weeds are plants that have escaped from cultivation, as the wild radish, which is troublesome in parts of New England; the wild carrot, which infests the fields in eastern New York;

and the live-forever, which thrives and multiplies under the plow and harrow. In my section an annoying weed is abutilon, or velvet-leaf, also called "old maid," which has fallen from the grace of the garden and followed the plow afield. It will manage to mature its seeds if not allowed to start till midsummer.

Of beautiful weeds quite a long list might be made without including any of the so-called wild flowers. A favorite of mine is the little moth mullein that blooms along the highway, and about the fields, and maybe upon the edge of the lawn, from midsummer till frost comes. In winter its slender stalk rises above the snow, bearing its round seedpods on its pin-like stems, and is pleasing even then. Its flowers are yellow or white, large, wheelshaped, and are borne vertically with filaments loaded with little tufts of violet wool. The plant has none of the coarse, hairy character of the common mullein. Our cone-flower, which one of our poets has called the "brown-eyed daisy," has a pleasing effect when in vast numbers they invade a meadow (if it is not your meadow), their dark brown centres or disks and their golden rays showing conspicuously.

Bidens, two-teeth, or "pitchforks," as the boys call them, are welcomed by the eye when in late summer they make the swamps and wet, waste places yellow with their blossoms.

A BUNCH OF HERBS

Vervain is a beautiful weed, especially the blue or purple variety. Its drooping knotted threads also make a pretty etching upon the winter snow.

Iron-weed, which looks like an overgrown aster, has the same intense purple-blue color, and a royal profusion of flowers. There are giants among the weeds, as well as dwarfs and pigmies. One of the giants is purple eupatorium, which sometimes carries its corymbs of flesh-colored flowers ten and twelve feet high. A pretty and curious little weed, sometimes found growing in the edge of the garden, is the clasping specularia, a relative of the harebell and of the European Venus's looking-glass. Its leaves are shell-shaped, and clasp the stalk so as to form little shallow cups. In the bottom of each cup three buds appear that never expand into flowers ; but when the top of the stalk is reached, one and sometimes two buds open a large, delicate purple-blue corolla. All the first-born of this plant are still-born, as it were ; only the latest, which spring from its summit, attain to perfect bloom. A weed which one ruthlessly demolishes when he finds it hiding from the plow amid the strawberries, or under the currant-bushes and grapevines, is the dandelion; yet who would banish it from the meadows or the lawns, where it copies in gold upon the green expanse the stars of the midnight sky ? After its first blooming comes its second and finer and more spiritual inflorescence, when its stalk, drop-

ping its more earthly and carnal flower, shoots
upward, and is presently crowned by a globe of
the most delicate and aerial texture. It is like the
poet's dream, which succeeds his rank and golden
youth. This globe is a fleet of a hundred fairy
balloons, each one of which bears a seed which it
is destined to drop far from the parent source.

Most weeds have their uses; they are not wholly
malevolent. Emerson says a weed is a plant whose
virtues we have not yet discovered ; but the wild
creatures discover their virtues if we do not. The
bumblebee has discovered that the hateful toad-
flax, which nothing will eat, and which in some
soils will run out the grass has honey at its heart.
Narrow-leaved plantain is readily eaten by cattle,
and the honey-bee gathers much pollen from it.
The ox-eye daisy makes a fair quality of hay if
cut before it gets ripe. The cows will eat the leaves
of the burdock and the stinging nettles of the woods.
But what cannot a cow's tongue stand ? She will
crop the poison ivy with impunity, and I think
would eat thistles if she found them growing in the
garden. Leeks and garlics are readily eaten by
cattle in the spring, and are said to be medicinal
to them. Weeds that yield neither pasturage for
bee nor herd yet afford seeds to the fall and winter
birds. This is true of most of the obnoxious weeds
of the garden, and of thistles. The wild lettuce
yields down for the hummingbird's nest, and the

flowers of whiteweed are used by the kingbird and cedar-bird.

Yet it is pleasant to remember that, in our climate, there are no weeds so persistent and lasting and universal as grass. Grass is the natural covering of the fields. There are but four weeds that I know of — milkweed, live-forever, Canada thistle, and toad-flax — that it will not run out in a good soil. We crop it and mow it year after year ; and yet, if the season favors, it is sure to come again. Fields that have never known the plow, and never been seeded by man, are yet covered with grass. And in human nature, too, weeds are by no means in the ascendant, troublesome as they are. The good green grass of love and truthfulness and common sense is more universal, and crowds the idle weeds to the wall.

But weeds have this virtue; they are not easily discouraged ; they never lose heart entirely ; they die game. If they cannot have the best, they will take up with the poorest ; if fortune is unkind to them to-day, they hope for better luck to-morrow; if they cannot lord it over a corn-hill, they will sit humbly at its foot and accept what comes ; in all cases they make the most of their opportunities.

A SHARP LOOKOUT

*John Burroughs with John Muir
in Yosemite National Park in 1909*

EDITOR'S NOTE

On March 1, 1882, after spending two weeks completing an essay called "Signs and Seasons," Burroughs recorded in his journal, "Writing is like fishing: you do not know that there are fish in the hole till you have caught them. I did not know there was an article in me on this subject till I fished it out." Originally published in *Century Magazine* in March 1883, the essay was revised and renamed "A Sharp Lookout" and printed as the initial chapter of the volume *Signs and Seasons,* published in 1886. In "A Sharp Lookout" Burroughs defines the methods that have come to be identified with him as a reliable observer in the natural world. He is both passive and active, a careful observer and questioner of the weather, trees, and insects close to home. He corrects his own earlier misobservations as readily as he corrects the natural history of the ancients and the folk wisdom of the almanac. He recognizes the links between the methods of poetry and science in the ways they process the hints and half-truths of nature, and he finds room to admire the modern scientist and the

331

EDITOR'S NOTE

ancient poet as well as the South American *rastreador*. Most important, though, is the "home feeling" he claims a reliable observer must develop so that "one's own landscape comes in time to be a sort of outlying part of himself; . . . cut those trees, and he bleeds; mar those hills, and he suffers. . . . Nature comes home to one most when he is at home."

F.B.

I

A SHARP LOOKOUT

ONE has only to sit down in the woods or the
fields, or by the shore of the river or the lake,
and nearly everything of interest will come round
to him, — the birds, the animals, the insects; and
presently, after his eye has got accustomed to the
place, and to the light and shade, he will probably
see some plant or flower that he has sought in vain,
and that is a pleasant surprise to him. So, on a
large scale, the student and lover of nature has this
advantage over people who gad up and down the
world, seeking some novelty or excitement; he has
only to stay at home and see the procession pass.
The great globe swings around to him like a revolv-
ing showcase; the change of the seasons is like the
passage of strange and new countries; the zones of
the earth, with all their beauties and marvels, pass
one's door, and linger long in the passing. What a
voyage is this we make without leaving for a night
our own fireside! St. Pierre well says that a sense
of the power and mystery of nature shall spring up

as fully in one's heart after he has made the circuit
of his own field as after returning from a voyage
round the world. I sit here amid the junipers of the
Hudson, with purpose every year to go to Florida,
or to the West Indies, or to the Pacific coast, yet the
seasons pass and I am still loitering, with a half-
defined suspicion, perhaps, that, if I remain quiet
and keep a sharp lookout, these countries will come
to me. I may stick it out yet, and not miss much
after all. The great trouble is for Mohammed to
know when the mountain really comes to him.
Sometimes a rabbit or a jay or a little warbler
brings the woods to my door. A loon on the river,
and the Canada lakes are here; the sea-gulls and the
fish hawk bring the sea; the call of the wild gander
at night, what does it suggest? and the eagle flapping by, or floating along on a raft of ice, does not
he bring the mountain? One spring morning five
swans flew above my barn in single file, going northward, — an express train bound for Labrador. It
was a more exhilarating sight than if I had seen
them in their native haunts. They made a breeze
in my mind, like a noble passage in a poem. How
gently their great wings flapped; how easy to fly
when spring gives the impulse! On another occasion I saw a line of fowls, probably swans, going
northward, at such a height that they appeared like
a faint, waving black line against the sky. They
must have been at an altitude of two or three miles.

A SHARP LOOKOUT

I was looking intently at the clouds to see which way they moved, when the birds came into my field of vision. I should never have seen them had they not crossed the precise spot upon which my eye was fixed. As it was near sundown, they were probably launched for an all-night pull. They were going with great speed, and as they swayed a little this way and that, they suggested a slender, all but invisible, aerial serpent cleaving the ether. What a highway was pointed out up there! — an easy grade from the Gulf to Hudson's Bay.

Then the typical spring and summer and autumn days, of all shades and complexions, — one cannot afford to miss any of them; and when looked out upon from one's own spot of earth, how much more beautiful and significant they are! Nature comes home to one most when he is at home; the stranger and traveler finds her a stranger and traveler also. One's own landscape comes in time to be a sort of outlying part of himself; he has sowed himself broadcast upon it, and it reflects his own moods and feelings; he is sensitive to the verge of the horizon : cut those trees, and he bleeds ; mar those hills, and he suffers. How has the farmer planted himself in his fields; builded himself into his stone walls, and evoked the sympathy of the hills by his struggle! This home feeling, this domestication of nature, is important to the observer. This is the bird-lime with which he catches the bird; this is the private door that admits him behind the scenes. This is one source of Gilbert White's charm, and of the charm of Thoreau's " Walden."

The birds that come about one's door in winter, or that build in his trees in summer, what a peculiar interest they have! What crop have I sowed in Florida or in California, that I should go there to reap? I should be only a visitor, or formal caller upon nature, and the family would all wear masks. No; the place to observe nature is where you are; the walk to take to-day is the walk you took yesterday. You will not find just the same things: both the observed and the observer have changed; the ship is on another tack in both cases.

I shall probably never see another just such day as yesterday was, because one can never exactly repeat his observation, — cannot turn the leaf of the book of life backward, — and because each day has characteristics of its own. This was a typical March day, clear, dry, hard, and windy, the river rumpled and crumpled, the sky intense, distant objects strangely near; a day full of strong light, unusual; an extraordinary lightness and clearness all around the horizon, as if there were a diurnal aurora streaming up and burning through the sunlight; smoke from the first spring fires rising up in various directions; a day that winnowed the air, and left no film in the sky. At night, how the big March bellows did work! Venus was like a great lamp in the sky. The stars all seemed brighter than usual, as if the wind blew them up like burning coals. Venus actually seemed to flare in the wind.

Each day foretells the next, if one could read the signs; to-day is the progenitor of to-morrow. When the atmosphere is telescopic, and distant objects stand out unusually clear and sharp, a storm is near. We are on the crest of the wave, and the depression follows quickly. It often happens that clouds are not so indicative of a storm as the total absence of clouds. In this state of the atmosphere the stars are unusually numerous and bright at night, which is also a bad omen.

I find this observation confirmed by Humboldt. "It appears," he says, "that the transparency of the air is prodigiously increased when a certain quantity of water is uniformly diffused through it." Again, he says that the mountaineers of the Alps "predict a change of weather when, the air being calm, the Alps covered with perpetual snow seem on a sudden to be nearer the observer, and their outlines are marked with great distinctness on the azure sky." He further observes that the same condition of the atmosphere renders distant sounds more audible.

There is one redness in the east in the morning that means storm, another that means wind. The former is broad, deep, and angry; the clouds look like a huge bed of burning coals just raked open ;

the latter is softer, more vapory, and more widely extended. Just at the point where the sun is going to rise, and some minutes in advance of his coming, there sometimes rises straight upward a rosy column; it is like a shaft of deeply dyed vapor, blending with and yet partly separated from the clouds, and the base of which presently comes to glow like the sun itself. The day that follows is pretty certain to be very windy. At other times the under sides of the eastern clouds are all turned to pink or rose-colored wool; the transformation extends until nearly the whole sky flushes, even the west glowing slightly; the sign is always to be interpreted as meaning fair weather.

The approach of great storms is seldom heralded by any striking or unusual phenomenon. The real weather gods are free from brag and bluster; but the sham gods fill the sky with portentous signs and omens. I recall one 5th of March as a day that would have filled the ancient observers with dreadful forebodings. At ten o'clock the sun was attended by four extraordinary sun-dogs. A large bright halo encompassed him, on the top of which the segment of a larger circle rested, forming a sort of heavy brilliant crown. At the bottom of the circle, and depending from it, was a mass of soft, glowing, iridescent vapor. On either side, like fragments of the larger circle, were two brilliant arcs. Altogether, it was the most portentous storm-breeding sun I ever beheld. In a dark hemlock wood in a valley, the owls were hooting ominously, and the crows dismally cawing. Before night the storm set in, a little sleet and rain of a few hours' duration, insignificant enough compared with the signs and wonders that preceded it.

To what extent the birds or animals can foretell the weather is uncertain. When the swallows are seen hawking very high, it is a good indication; the insects upon which they feed venture up there only in the most auspicious weather. Yet bees will continue to leave the hive when a storm is imminent. I am told that one of the most reliable weather signs they have down in Texas is afforded by the ants. The ants bring their eggs up out of their underground retreats and expose them to the warmth of the sun to be hatched. When they are seen carrying them in again in great haste, though there be not a cloud in the sky, your walk or your drive must be postponed: a storm is at hand. There is a passage in Virgil that is doubtless intended to embody a similar observation, though none of his translators seem to have hit its meaning accurately: —

" Sæpius et tectis penetralibus extulit ova
    Angustum formica terens iter;' '

" Often also has the pismire making a narrow road brought forth her eggs out of the hidden recesses," is the literal translation of old John Martyn.

" Also the ant, incessantly traveling
  The same straight way with the eggs of her hidden
      store,"

is one of the latest metrical translations. Dryden
has it: —

  " The careful ant her secret cell forsakes
  And drags her eggs along the narrow tracks,"

which comes nearer to the fact. When a storm is
coming, Virgil also makes his swallows skim low
about the lake, which agrees with the observation
above.

The critical moments of the day as regards the
weather are at sunrise and sunset. A clear sunset
is always a good sign; an obscured sun, just at the
moment of going down after a bright day, bodes
storm. There is much truth, too, in the saying
that if it rain before seven, it will clear before
eleven. Nine times in ten it will turn out thus.
The best time for it to begin to rain or snow, if it
wants to hold out, is about mid-forenoon. The
great storms usually begin at this time. On all
occasions the weather is very sure to declare itself
before eleven o'clock. If you are going on a pic-
nic, or are going to start on a journey, and the
morning is unsettled, wait till ten and one half
o'clock, and you will know what the remainder
of the day will be. Midday clouds and afternoon
clouds, except in the season of thunderstorms, are

usually harmless idlers and vagabonds. But more to be relied on than any obvious sign is that subtle perception of the condition of the weather which a man has who spends much of his time in the open air. He can hardly tell how he knows it is going to rain; he hits the fact as an Indian does the mark with his arrow, without calculating and by a kind of sure instinct. As you read a man's purpose in his face, so you learn to read the purpose of the weather in the face of the day.

In observing the weather, however, as in the diagnosis of disease, the diathesis is all-important. All signs fail in a drought, because the predisposition, the diathesis, is so strongly toward fair weather; and the opposite signs fail during a wet spell, because nature is caught in the other rut.

Observe the lilies of the field. Sir John Lubbock says the dandelion lowers itself after flowering, and lies close to the ground while it is maturing its seed, and then rises up. It is true that the dandelion lowers itself after flowering, retires from society, as it were, and meditates in seclusion; but after it lifts itself up again, the stalk begins anew to grow, it lengthens daily, keeping just above the grass till the fruit is ripened, and the little globe of silvery down is carried many inches higher than was the ring of golden flowers. And the reason is obvious. The plant depends upon the wind to scatter its seeds; every one of these little vessels spreads a sail to the breeze, and it is necessary that they be launched above the grass and weeds, amid which they would be caught and held did the stalk not continue to grow and outstrip the rival vegetation. It is a curious instance of foresight in a weed.

I wish I could read as clearly this puzzle of the button-balls (American plane-tree). Why has Nature taken such particular pains to keep these balls hanging to the parent tree intact till spring? What secret of hers has she buttoned in so securely? for these buttons will not come off. The wind cannot twist them off, nor warm nor wet hasten or retard them. The stem, or peduncle, by which the ball is held in the fall and winter, breaks up into a dozen or more threads or strands, that are stronger than those of hemp. When twisted tightly they make a little cord that I find impossible to break with my hands. Had they been longer, the Indian would surely have used them to make his bow-strings and all the other strings he required. One could hang himself with a small cord of them. (In South America, Humboldt saw excellent cordage made by the Indians from the petioles of the Chiquichiqui palm.) Nature has determined that these buttons should stay on. In order that the seeds of this tree may germinate, it is probably necessary that they be kept dry during the winter, and reach the ground after the season of warmth and moisture is fully established. In May, just as the leaves and the new balls are emerging, at the touch of a warm, moist south wind, these spherical packages suddenly go to pieces — explode, in fact, like tiny bombshells that were fused to carry to this point — and scatter their seeds to the four winds. They yield at the same time a fine pollen-like dust that one would suspect played some part in fertilizing the new balls, did not botany teach him otherwise. At any rate, it is the only deciduous tree I know of that does not let go the old seed till the new is well on the way. It is plain why the sugar-berry-tree or lotus holds its drupes all winter: it is in order that the birds may come and sow the seed. The berries are like small gravel-stones with a sugar coating, and a bird will not eat them till he is pretty hard pressed, but in late fall and winter the robins, cedar-birds, and bluebirds devour them readily, and of course lend their wings to scatter the seed far and wide. The same is true of juniper-berries, and the fruit of the bitter-sweet.

In certain other cases where the fruit tends to hang on during the winter, as with the bladder-nut and the honey-locust, it is probably because the frost and the perpetual moisture of the ground would rot or kill the germ. To beechnuts, chestnuts, and acorns the moisture of the ground and the covering of leaves seem congenial, though too much warmth and moisture often cause the acorns to germinate prematurely. I have found the ground under the oaks in December covered with nuts, all anchored to the earth by purple sprouts. But the winter which follows such untimely growths generally proves fatal to them.

One must always cross-question nature if he would get at the truth, and he will not get at it then unless he frames his questions with great skill. Most persons are unreliable observers because they put only leading questions, or vague questions.

Perhaps there is nothing in the operations of nature to which we can properly apply the term intelligence, yet there are many things that at first sight look like it. Place a tree or plant in an unusual position and it will prove itself equal to the occasion, and behave in an unusual manner; it will show original resources; it will seem to try intelligently to master the difficulties. Up by Furlow Lake, where I was camping out, a young hemlock had become established upon the end of a large and partly decayed log that reached many feet out into the lake. The young tree was eight or nine feet high; it had sent its roots down into the log and clasped it around on the outside, and had apparently discovered that there was water instead of soil immediately beneath it, and that its sustenance must be sought elsewhere and that quickly. Accordingly it had started one large root, by far the largest of all, for the shore along the top of the log. This root, when I saw the tree, was six or seven feet long, and had bridged more than half the distance that separated the tree from the land.

Was this a kind of intelligence? If the shore had lain in the other direction, no doubt at all but the root would have started for the other side. I know a yellow pine that stands on the side of a steep hill. To make its position more secure, it has thrown out a large root at right angles with its stem directly into the bank above it, which acts as a stay or guy-rope. It was positively the best thing the tree could do. The earth has washed away so that the root where it leaves the tree is two feet above the surface of the soil.

Yet both these cases are easily explained, and without attributing any power of choice, or act of intelligent selection, to the trees. In the case of the little hemlock upon the partly submerged log, roots were probably thrown out equally in all directions; on all sides but one they reached the water and stopped growing; the water checked them; but on the land side, the root on the top of the log, not meeting with any obstacle of the kind, kept on growing, and thus pushing its way toward the shore. It was a case of survival, not of the fittest, but of that which the situation favored, — the fittest with reference to position.

So with the pine-tree on the side of the hill. It probably started its roots in all directions, but only the one on the upper side survived and matured. Those on the lower side finally perished, and others lower down took their places. Thus the whole life upon the globe, as we see it, is the result of this blind groping and putting forth of Nature in every direction, with failure of some of her ventures and the success of others, the circumstances, the environments, supplying the checks and supplying the stimulus, the seed falling upon the barren places just the same as upon the fertile. No discrimination on the part of Nature that we can express in the terms of our own consciousness, but ceaseless experiments in every possible direction. The only thing inexplicable is the inherent impulse to experiment, the original push, the principle of Life.

The good observer of nature holds his eye long and firmly to the point, as one does when looking at a puzzle picture, and will not be baffled. The cat catches the mouse, not merely because she watches for him, but because she is armed to catch him and is quick. So the observer finally gets the fact, not only because he has patience, but because his eye is sharp and his inference swift. Many a shrewd old farmer looks upon the milky way as a kind of weathercock, and will tell you that the way it points at night indicates the direction of the wind the following day. So, also, every new moon is either a dry moon or a wet moon, dry if a powder-horn would hang upon the lower limb, wet if it would not; forgetting the fact that, as a rule, when it is dry in one part of the continent it is wet in some other part, and *vice versa*. When he kills his hogs in the fall, if the pork be very hard and solid, he predicts a severe winter; if soft and loose, the opposite; again overlooking the fact that the kind of food and the temperature of the fall make the pork hard or make it soft. So with a hundred other signs, all the result of hasty and incomplete observations.

One season, the last day of December was very warm. The bees were out of the hive, and there was no frost in the air or in the ground. I was walking in the woods, when as I paused in the shade of a hemlock-tree I heard a sound proceed from beneath the wet leaves on the ground but a few feet from me that suggested a frog. Following it cautiously up, I at last determined upon the exact spot whence the sound issued; lifting up the thick layer of leaves, there sat a frog — the wood frog, one of the first to appear in the marshes in spring, and which I have elsewhere called the "clucking frog" —in a little excavation in the surface of the leaf mould. As it sat there, the top of its back was level with the surface of the ground. This, then, was its hibernaculum; here it was prepared to pass the winter, with only a coverlid of wet matted leaves between it and zero weather. Forthwith I set up as a prophet of warm weather, and among other things predicted a failure of the ice crop on the river; which, indeed, others, who had not heard frogs croak on the 31st of December, had also begun to predict. Surely, I thought, this frog knows what it is about; here is the wisdom of nature; it would have gone deeper into the ground than that if a severe winter was approaching; so I was not anxious about my coal-bin, nor disturbed by longings for Florida. But what a winter followed, the winter of 1885, when the Hudson became coated with ice nearly two feet thick, and when March was as cold as January! I thought of my frog under the hemlock and wondered how it was faring. So one day the latter part of March, when the snow was gone, and there was a feeling of spring in the air, I turned aside in my walk to investigate it. The matted leaves were still frozen hard, but I succeeded in lifting them up and exposing the frog. There it sat as fresh and unscathed as in the fall. The ground beneath and all about it was still frozen like a rock, but apparently it had some means of its own of resisting the frost. It winked and bowed its head when I touched it, but did not seem inclined to leave its retreat. Some days later, after the frost was nearly all out of the ground, I passed that way, and found my frog had come out of its seclusion and was resting amid the dry leaves. There was not much jump in it yet, but its color was growing lighter. A few more warm days, and its fellows, and doubtless itself too, were croaking and gamboling in the marshes.

This incident convinced me of two things; namely, that frogs know no more about the coming weather than we do, and that they do not retreat as deep into the ground to pass the winter as has been supposed. I used to think the muskrats could fore-tell an early and a severe winter, and have so writ-ten. But I am now convinced they cannot; they know as little about it as I do. Sometimes on an early and severe frost they seem to get alarmed and go to building their houses, but usually they seem to build early or late, high or low, just as the whim takes them.

In most of the operations of nature there is at least one unknown quantity; to find the exact value of this unknown factor is not so easy. The wool of the sheep, the fur of the animals, the feathers of the fowls, the husks of the maize, why are they thicker some seasons than others; what is the value of the unknown quantity here? Does it indicate a severe winter approaching? Only observations extending over a series of years could determine the point. How much patient observation it takes to settle many of the facts in the lives of the birds, ani-mals, and insects! Gilbert White was all his life trying to determine whether or not swallows passed the winter in a torpid state in the mud at the bot-tom of ponds and marshes, and he died ignorant of the truth that they do not. Do honey-bees injure the grape and other fruits by puncturing the skin for the juice? The most patient watching by many skilled eyes all over the country has not yet settled the point. For my own part, I am convinced that they do not. The honey-bee is not the rough-and-ready freebooter that the wasp and the bumblebee are; she has somewhat of feminine timidity, and leaves the first rude assaults to them. I knew the honey-bee was very fond of the locust blossoms, and that the trees hummed like a hive in the height of their flowering, but I did not know that the bumblebee was ever the sapper and miner that went ahead in this enterprise, till one day I placed myself amid the foliage of a locust and saw him savagely bite through the shank of the flower and extract the nectar, followed by a honey-bee that in every instance searched for this opening, and probed long and carefully for the leavings of her burly purveyor. The bumblebee rifles the dicentra and the columbine of their treasures in the same manner, namely, by slitting their pockets from the outside, and the honey-bee gleans after him, taking the small change he leaves. In the case of the locust, however, she usually obtains the honey without the aid of the larger bee.

Speaking of the honey-bee reminds me that the subtle and sleight-of-hand manner in which she fills her baskets with pollen and propolis is character istic of much of Nature's doings. See the bee going from flower to flower with the golden pellets on her thighs, slowly and mysteriously increasing in size. If the miller were to take the toll of the grist he grinds by gathering the particles of flour from his coat and hat, as he moved rapidly about, or catching them in his pockets, he would be doing pretty nearly what the bee does. The little miller dusts herself with the pollen of the flower, and then, while on the wing, brushes it off with the fine brush on certain of her feet, and by some jugglery or other catches it in her pollen basket. One needs to look long and intently to see through the trick. Pliny says they fill their baskets with their fore feet, and that they fill their fore feet with their trunks, but it is a much more subtle operation than this. I have seen the bees come to a meal barrel in early spring, and to a pile of hardwood sawdust before there was yet anything in nature for them to work upon, and, having dusted their coats with the finer particles of the meal or the sawdust, hover on the wing above the mass till the little legerdemain feat was performed. Nature fills her baskets by the same sleight-of-hand, and the observer must be on the alert who would possess her secret. If the ancients had looked a little closer and sharper, would they ever have believed in spontaneous generation in the superficial way in which they did ; that maggots, for instance, were gen-

erated spontaneously in putrid flesh? Could they
not see the spawn of the blow-flies? Or, if Virgil
had been a real observer of the bees, would he ever
have credited, as he certainly appears to do, the
fable of bees originating from the carcass of a steer?
or that on windy days they carried little stones for
ballast? or that two hostile swarms fought each
other in the air? Indeed, the ignorance, or the
false science, of the ancient observers, with regard
to the whole subject of bees, is most remarkable;
not false science merely with regard to their more
hidden operations, but with regard to that which
is open and patent to all who have eyes in their
heads, and have ever had to do with them. And
Pliny names authors who had devoted their whole
lives to the study of the subject.

But the ancients, like women and children, were
not accurate observers. Just at the critical moment
their eyes were unsteady, or their fancy, or their
credulity, or their impatience, got the better of
them, so that their science was half fact and half
fable. Thus, for instance, because the young cuckoo
at times appeared to take the head of its small
foster mother quite into its mouth while receiving
its food, they believed that it finally devoured her.
Pliny, who embodied the science of his times in
his natural history, says of the wasp that it carries
spiders to its nest, and then sits upon them until
it hatches its young from them. A little careful

observation would have shown him that this was only a half truth; that the whole truth was, that the spiders were entombed with the egg of the wasp to serve as food for the young when the egg had hatched.

What curious questions Plutarch discusses, as, for instance, "What is the reason that a bucket of water drawn out of a well, if it stands all night in the air that is in the well, is more cold in the morning than the rest of the water?" He could probably have given many reasons why "a watched pot never boils." The ancients, the same author says, held that the bodies of those killed by lightning never putrefy; that the sight of a ram quiets an enraged elephant; that a viper will lie stock-still if touched by a beechen leaf; that a wild bull grows tame if bound with the twigs of a fig-tree; that a hen purifies herself with straw after she has laid an egg; that the deer buries his cast-off horns; and that a goat stops the whole herd by holding a branch of the sea-holly in his mouth. They sought to account for such things without stopping to ask, Are they true? Nature was too novel, or else too fearful, to them to be deliberately pursued and hunted down. Their youthful joy in her, or their dread and awe in her presence, may be better than our scientific satisfaction, or cool wonder, or our vague, mysterious sense of "something far more deeply interfused;" yet we cannot change with them if we would, and I, for one, would not if I could. Science does not mar nature. The railroad, Thoreau found, after all, to be about the wildest road he knew of, and the telegraph wires the best æolian harp out of doors. Study of nature deepens the mystery and the charm because it removes the horizon farther off. We cease to fear, perhaps, but how can one cease to marvel and to love?

The fields and woods and waters about one are a book from which he may draw exhaustless entertainment, if he will. One must not only learn the writing, he must translate the language, the signs, and the hieroglyphics. It is a very quaint and elliptical writing, and much must be supplied by the wit of the translator. At any rate, the lesson is to be well conned. Gilbert White said that that locality would be found the richest in zoölogical or botanical specimens which was most thoroughly examined. For more than forty years he studied the ornithology of his district without exhausting the subject. I thought I knew my own tramping-ground pretty well, but one April day, when I looked a little closer than usual into a small semi-stagnant lakelet where I had peered a hundred times before, I suddenly discovered scores of little creatures that were as new to me as so many nymphs would have been. They were partly fish-shaped, from an inch to an inch and a half long, semi-transparent, with a dark brownish line visible the entire length of them (apparently the thread upon which the life of the animal hung, and by which its all but impalpable frame was held together), and suspending themselves in the water, or impelling themselves swiftly forward by means of a double row of fine, waving, hair-like appendages, that arose from what appeared to be the back, — a kind of undulating, pappus-like wings. What was it? I did not know. None of my friends or scientific acquaintances knew. I wrote to a learned man, an authority upon fish, describing the creature as well as I could. He replied that it was only a familiar species of phyllopodous crustacean, known as *Eubranchipus vernalis.*

I remember that our guide in the Maine woods, seeing I had names of my own for some of the plants, would often ask me the name of this and that flower for which he had no word ; and that when I could recall the full Latin term, it seemed overwhelmingly convincing and satisfying to him. It was evidently a relief to know that these obscure plants of his native heath had been found worthy of a learned name, and that the Maine woods were not so uncivil and outlandish as they might at first seem: it was a comfort to him to know that he did not live beyond the reach of botany. In like manner I found satisfaction in knowing that my novel fish had been recognized and worthily named; the title conferred a new dignity at once; but when the learned man added that it was familiarly called the " fairy shrimp," I felt a deeper pleasure. Fairy-like it certainly was, in its aerial, unsubstantial look, and in its delicate, down-like means of loco-motion; but the large head, with its curious folds, and its eyes standing out in relief, as if on the heads of two pins, was gnome-like. Probably the fairy wore a mask, and wanted to appear terrible to human eyes. Then the creatures had sprung out of the earth as by magic. I found some in a fur-row in a plowed field that had encroached upon a swamp. In the fall the plow had been there, and had turned up only the moist earth ; now a little water was standing there, from which the April sunbeams had invoked these airy, fairy creatures. They belong to the crustaceans, but apparently no creature has so thin or impalpable a crust; you can almost see through them ; certainly you can see what they have had for dinner, if they have eaten substantial food.

All we know about the private and essential natural history of the bees, the birds, the fishes, the animals, the plants, is the result of close, pa-tient, quick-witted observation. Yet Nature will often elude one for all his pains and alertness. Thoreau, as revealed in his journal, was for years trying to settle in his own mind what was the first thing that stirred in spring, after the severe New England winter, — in what was the first sign or

pulse of returning life manifest; and he never seems
to have been quite sure. He could not get his salt
on the tail of this bird. He dug into the swamps,
he peered into the water, he felt with benumbed
hands for the radical leaves of the plants under the
snow; he inspected the buds on the willows, the
catkins on the alders; he went out before daylight
of a March morning and remained out after dark;
he watched the lichens and mosses on the rocks;
he listened for the birds; he was on the alert for
the first frog (" Can you be absolutely sure," he
says, " that you have heard the first frog that
croaked in the township ? "); he stuck a pin here
and he stuck a pin there, and there, and still he
could not satisfy himself. Nor can any one. Life
appears to start in several things simultaneously.
Of a warm thawy day in February the snow is
suddenly covered with myriads of snow fleas looking like black, new powder just spilled there. Or
you may see a winged insect in the air. On the
selfsame day the grass in the spring run and the
catkins on the alders will have started a little; and
if you look sharply, while passing along some sheltered nook or grassy slope where the sunshine lies
warm on the bare ground, you will probably see a
grasshopper or two. The grass hatches out under
the snow, and why should not the grasshopper ?
At any rate, a few such hardy specimens may be
found in the latter part of our milder winters

wherever the sun has uncovered a sheltered bit of grass for a few days, even after a night of ten or twelve degrees of frost. Take them in the shade, and let them freeze stiff as pokers, and when thawed out again they will hop briskly. And yet, if a poet were to put grasshoppers in his winter poem, we should require pretty full specifications of him, or else fur to clothe them with. Nature will not be cornered, yet she does many things in a corner and surreptitiously. She is all things to all men; she has whole truths, half truths, and quarter truths, if not still smaller fractions. The careful observer finds this out sooner or later. Old fox-hunters will tell you, on the evidence of their own eyes, that there is a black fox and a silver-gray fox, two species, but there are not ; the black fox is black when coming toward you or running from you, and silver gray at point-blank view, when the eye penetrates the fur; each separate hair is gray the first half and black the last. This is a sample of Nature's half truths.

Which are our sweet-scented wild flowers ? Put your nose to every flower you pluck, and you will be surprised how your list will swell the more you smell. I plucked some wild blue violets one day, the *ovata* variety of the *sagittata*, that had a faint perfume of sweet clover, but I never could find another that had any odor. A pupil disputed with his teacher about the hepatica, claiming in opposition that it was sweet-scented. Some hepaticas are sweet-scented and some are not, and the perfume is stronger some seasons than others. After the unusually severe winter of 1880–81, the variety of hepatica called the sharp-lobed was markedly sweet in nearly every one of the hundreds of specimens I examined. A handful of them exhaled a most delicious perfume. The white ones that season were largely in the ascendant; and probably the white specimens of both varieties, one season with another, will oftenest prove sweet-scented. Darwin says a considerably larger proportion of white flowers are sweet-scented than of any other color. The only sweet violets I can depend upon are white, *Viola blanda* and *Viola Canadensis*, and white largely predominates among our other odorous wild flowers. All the fruit trees have white or pinkish blossoms. I recall no native blue flower of New York or New England that is fragrant except in the rare case of the arrow-leaved violet, above referred to. The earliest yellow flowers, like the dandelion and yellow violets, are not fragrant. Later in the season yellow is frequently accompanied with fragrance, as in the evening primrose, the yellow lady's-slipper, horned bladderwort, and others.

My readers probably remember that on a former occasion I have mildly taken the poet Bryant to task for leading his readers to infer that the early yellow violet is sweet-scented. In view of the capriciousness of the perfume of certain of our wild flowers, I have during the past few years tried industriously to convict myself of error in respect to this flower. The round-leaved yellow violet was one of the earliest and most abundant wild flowers in the woods where my youth was passed, and whither I still make annual pilgrimages. I have pursued it on mountains and in lowlands, in " beechen woods " and amid the hemlocks ; and while, with respect to its earliness, it overtakes the hepatica in the latter part of April, as do also the dog's-tooth violet and the claytonia, yet the first hepaticas, where the two plants grow side by side, bloom about a week before the first violet. And I have yet to find one that has an odor that could be called a perfume. A handful of them, indeed, has a faint, bitterish smell, not unlike that of the dandelion in quality; but if every flower that has a smell is sweet-scented, then every bird that makes a noise is a songster.

On the occasion above referred to, I also dissented from Lowell's statement, in " Al Fresco," that in early summer the dandelion blooms, in general, with the buttercup and the clover. I am aware that such criticism of the poets is small game, and not worth the powder. General truth, and not specific fact, is what we are to expect of the poets. Bryant's " Yellow Violet " poem is tender and appropriate, and such as only a real lover and observer of nature could feel or express; and Lowell's "Al Fresco" is full of the luxurious feeling of early summer, and this is, of course, the main thing; a good reader cares for little else; I care for little else myself. But when you take your coin to the assay office, it must be weighed and tested, and in the comments referred to I (unwisely, perhaps) sought to smelt this gold of the poets in the naturalist's pot, to see what alloy of error I could detect in it. Were the poems true to their last word? They were not, and much subsequent investigation has only confirmed my first analysis. The general truth is on my side, and the specific fact, if such exists in this case, on the side of the poets. It is possible that there may be a fragrant yellow violet, as an exceptional occurrence, like that of the sweet-scented, arrow-leaved species above referred to, and that in some locality it may have bloomed before the hepatica; also that Lowell may have seen a belated dandelion or two in June, amid the clover and the buttercups; but, if so, they were the exception, and not the rule, — the specific or accidental fact, and not the general truth.

Dogmatism about nature, or about anything else, very often turns out to be an ungrateful cur that bites the hand that reared it. I speak from experience. I was once quite certain that the honey-bee did not work upon the blossoms of the trailing arbutus, but while walking in the woods one April

day I came upon a spot of arbutus swarming with
honey-bees. They were so eager for it that they
crawled under the leaves and the moss to get at the
blossoms, and refused on the instant the hive-honey
which I happened to have with me, and which
I offered them. I had had this flower under observation more than twenty years, and had never before seen it visited by honey-bees. The same season
I saw them for the first time working upon the
flower of bloodroot and of adder's-tongue. Hence
I would not undertake to say again what flowers
bees do not work upon. Virgil implies that they
work upon the violet, and for aught I know they
may. I have seen them very busy on the blossoms
of the white oak, though this is not considered a
honey or pollen yielding tree. From the smooth
sumac they reap a harvest in midsummer, and in
March they get a good grist of pollen from the
skunk-cabbage.

I presume, however, it would be safe to say that
there is a species of smilax with an unsavory name
that the bee does not visit, *Smilax herbacea.* The
production of this plant is a curious freak of nature.
I find it growing along the fences where one would
look for wild roses or the sweetbrier ; its recurv-
ing or climbing stem, its glossy, deep-green, heart-
shaped leaves, its clustering umbels of small green-
ish yellow flowers, making it very pleasing to the
eye; but to examine it closely one must positively

hold his nose. It would be too cruel a joke to offer it to any person not acquainted with it to smell. It is like the vent of a charnel-house. It is first cousin to the trilliums, among the prettiest of our native wild flowers, and the same bad blood crops out in the purple trillium or birthroot.

Nature will include the disagreeable and repulsive also. I have seen the phallic fungus growing in June under a rosebush. There was the rose, and beneath it, springing from the same mould, was this diabolical offering to Priapus. With the perfume of the roses into the open window came the stench of this hideous parody, as if in mockery. I removed it, and another appeared in the same place shortly afterward. The earthman was rampant and insulting. Pan is not dead yet. At least he still makes a ghastly sign here and there in nature.

The good observer of nature exists in fragments, a trait here and a trait there. Each person sees what it concerns him to see. The fox-hunter knows pretty well the ways and habits of the fox, but on any other subject he is apt to mislead you. He comes to see only fox traits in whatever he looks upon. The bee-hunter will follow the bee, but lose the bird. The farmer notes what affects his crops and his earnings, and little else. Common people, St. Pierre says, observe without reasoning, and the learned reason without observing. If one could apply to the observation of nature the sense and skill of the South American *rastreador*, or trailer, how much he would track home! This man's eye, according to the accounts of travelers, is keener than a hound's scent. A fugitive can no more elude him than he can elude fate. His perceptions are said to be so keen that the displacement of a leaf or pebble, or the bending down of a spear of grass, or the removal of a little dust from the fence, is enough to give him the clew. He sees the half-obliterated footprints of a thief in the sand, and carries the impression in his eye till a year afterward, when he again detects the same footprint in the suburbs of a city, and the culprit is tracked home and caught. I knew a man blind from his youth who not only went about his own neighborhood without a guide, turning up to his neighbor's gate or door as unerringly as if he had the best of eyes, but who would go many miles on an errand to a new part of the country. He seemed to carry a map of the township in the bottom of his feet, a most minute and accurate survey. He never took the wrong road, and he knew the right house when he had reached it. He was a miller and fuller, and ran his mill at night while his sons ran it by day. He never made a mistake with his customers' bags or wool, knowing each man's by the sense of touch. He frightened a colored man whom he detected stealing, as if he had seen out of the back of his head. Such facts show one how delicate and sensitive a man's relation to outward nature through his bodily senses may become. Heighten it a little more, and he could forecast the weather and the seasons, and detect hidden springs and minerals. A good observer has something of this delicacy and quickness of perception. All the great poets and naturalists have it. Agassiz traces the glaciers like a *rastreador;* and Darwin misses no step that the slow but tireless gods of physical change have taken, no matter how they cross or retrace their course. In the obscure fish-worm he sees an agent that has kneaded and leavened the soil like giant hands.

One secret of success in observing nature is capacity to take a hint; a hair may show where a lion is hid. One must put this and that together, and value bits and shreds. Much alloy exists with the truth. The gold of nature does not look like gold at the first glance. It must be smelted and refined in the mind of the observer. And one must crush mountains of quartz and wash hills of sand to get it. To know the indications is the main matter. People who do not know the secret are eager to take a walk with the observer to find where the mine is that contains such nuggets, little knowing that his ore-bed is but a gravel-heap to them. How insignificant appear most of the facts which one sees in his walks, in the life of the birds, the flowers, the animals, or in the phases of

the landscape, or the look of the sky!—insignifi
cant until they are put through some mental o
emotional process and their true value appears.
The diamond looks like a pebble until it is cut.
One goes to Nature only for hints and half truths.
Her facts are crude until you have absorbed them
or translated them. Then the ideal steals in and
lends a charm in spite of one. It is not so much
what we see as what the thing seen suggests. We
all see about the same; to one it means much, to
another little. A fact that has passed through the
mind of man, like lime or iron that has passed
through his blood, has some quality or property
superadded or brought out that it did not possess
before. You may go to the fields and the woods,
and gather fruit that is ripe for the palate without
any aid of yours, but you cannot do this in science
or in art. Here truth must be disentangled and
interpreted, — must be made in the image of man.
Hence all good observation is more or less a re-
fining and transmuting process, and the secret is to
know the crude material when you see it. I think
of Wordsworth's lines:—

> "The mighty world
> Of eye and ear, both what they half create, and what
> perceive;"

which is as true in the case of the naturalist as of
the poet; both "half create" the world they describe.

A SHARP LOOKOUT

Darwin does something to his facts as well as Tennyson to his. Before a fact can become poetry, it must pass through the heart or the imagination of the poet; before it can become science, it must pass through the understanding of the scientist. Or one may say, it is with the thoughts and half thoughts that the walker gathers in the woods and fields, as with the common weeds and coarser wild flowers which he plucks for a bouquet, — wild carrot, purple aster, moth mullein, sedge, grass, etc. : they look common and uninteresting enough there in the fields, but the moment he separates them from the tangled mass, and brings them indoors, and places them in a vase, say of some choice glass, amid artificial things, — behold, how beautiful! They have an added charm and significance at once; they are defined and identified, and what was common and familiar becomes unexpectedly attractive. The writer's style, the quality of mind he brings, is the vase in which his commonplace impressions and incidents are made to appear so beautiful and significant.

Man can have but one interest in nature, namely, to see himself reflected or interpreted there, and we quickly neglect both poet and philosopher who fail to satisfy, in some measure, this feeling.

THE TRAGEDIES OF THE NESTS

*Pointing out a junco's nest
by a mountain roadside*

EDITOR'S NOTE

"The Tragedies of the Nests" recounts Burroughs's seasonal observations of nesting birds during 1881. The essay was first published in *Century Magazine* in September 1883 and reprinted in the volume *Signs and Seasons* in 1886. Despite Burroughs's initial judgment of *Signs and Seasons* as "probably the least valuable of my books," it has become one of his most popular nature books, and Walt Whitman, objecting to Burroughs's depreciation of it, characterized the book as "so nice and fresh, like a new pot cheese in a clean napkin." Part of Burroughs's popular appeal lay in his personification of birds, but in contrast to other writers' popular and cozy scenes of birds as miniature human beings, Burroughs offers a "season of calamities, of violent deaths, of pillage and massacre." Behind his anthropomorphic tendencies lay a view of nature as blind and groping, disordered, and frequently ineffective. Even the instincts of birds and animals are not unerring. The powerful and elemental forces of necessity and accident evident in this essay

EDITOR'S NOTE

seemed to dominate Burroughs's attention during
this period. In preparing this essay for book publication, he sought to strengthen the ending by
bringing into sharper focus one of the predators—
the weasel—responsible for the havoc he had documented. After adding the concluding paragraphs
on the weasel, he ended the essay with an acknowl-
edgment to Darwin and the fundamental mysteri-
ousness of nature's processes.

F.B.

THE TRAGEDIES OF THE NESTS

THE life of the birds, especially of our migratory song-birds, is a series of adventures and of hairbreadth escapes by flood and field. Very few of them probably die a natural death, or even live out half their appointed days. The home instinct is strong in birds, as it is in most creatures; and I am convinced that every spring a large number of those which have survived the southern campaign return to their old haunts to breed. A Connecticut farmer took me out under his porch one April day, and showed me a phœbe-bird's nest six stories high. The same bird had no doubt returned year after year; and as there was room for only one nest upon her favorite shelf, she had each season reared a new superstructure upon the old as a foundation. I have heard of a white robin — an albino — that nested several years in succession in the suburbs of a Maryland city. A sparrow with a very marked peculiarity of song I have heard several seasons in my own locality. But the birds do not all live to return to their old haunts : the bobolinks and starlings run a gauntlet of fire from the Hudson to the Savannah, and the robins and meadowlarks and other song-birds are shot by boys and pot-hunters in great numbers, — to say nothing of their danger from hawks and owls. But of those that do return, what perils beset their nests, even in the most favored localities! The cabins of the early settlers, when the country was swarming with hostile Indians, were not surrounded by such dangers. The tender households of the birds are not only exposed to hostile Indians in the shape of cats and collectors, but to numerous murderous and bloodthirsty animals, against whom they have no defense but concealment. They lead the darkest kind of pioneer life, even in our gardens and orchards, and under the walls of our houses. Not a day or a night passes, from the time the eggs are laid till the young are flown, when the chances are not greatly in favor of the nest being rifled and its contents devoured, — by owls, skunks, minks, and coons at night, and by crows, jays, squirrels, weasels, snakes, and rats during the day. Infancy, we say, is hedged about by many perils; but the infancy of birds is cradled and pillowed in peril. An old Michigan settler told me that the first six children that were born to him died; malaria and teething invariably carried them off when they had reached a certain age; but other children were born, the country improved, and by and by the babies weathered the critical period, and the next six lived and grew up. The birds, too, would no doubt persevere six times and twice six times, if the season were long enough, and finally rear their family, but the waning summer cuts them short, and but few species have the heart and strength to make even the third trial.

The first nest-builders in spring, like the first settlers near hostile tribes, suffer the most casualties. A large proportion of the nests of April and May are destroyed; their enemies have been many months without eggs, and their appetites are keen for them. It is a time, too, when other food is scarce, and the crows and squirrels are hard put. But the second nests of June, and still more the nests of July and August, are seldom molested. It is rarely that the nest of the goldfinch or the cedar-bird is harried.

My neighborhood on the Hudson is perhaps exceptionally unfavorable as a breeding haunt for birds, owing to the abundance of fish crows and of red squirrels; and the season of which this chapter is mainly a chronicle, the season of 1881, seems to have been a black-letter one even for this place, for at least nine nests out of every ten that I observed during that spring and summer failed of their proper issue. From the first nest I noted, which was that of a bluebird, — built (very imprudently, I thought at the time) in a squirrel-hole in a decayed apple-tree, about the last of April, and which came to naught, even the mother bird, I suspect, perishing by a violent death,—to the last, which was that of a snowbird, observed in August, among the Catskills, deftly concealed in a mossy bank by the side of a road that skirted a wood, where the tall thimble blackberries grew in abundance, and from which the last young one was taken, when it was about half grown, by some nocturnal walker or daylight prowler, some untoward fate seemed hovering about them. It was a season of calamities, of violent deaths, of pillage and massacre, among our feathered neighbors. For the first time I noticed that the orioles were not safe in their strong pendent nests. Three broods were started in the apple-trees, only a few yards from the house, where, for several previous seasons, the birds had nested without molestation ; but this time the young were all destroyed when about half grown. Their chirping and chattering, which was so noticeable one day, suddenly ceased the next. The nests were probably plundered at night, and doubtless by the little red screech owl, which I know is a denizen of these old orchards, living in the deeper cavities of the trees. The owl could alight upon the top of the nest, and easily thrust his murderous claw down into its long pocket and seize the young and draw them forth. The tragedy of one of the nests was heightened, or at least made more palpable, by one of the half-fledged birds, either in its attempt to escape or while in the clutches of the enemy, being caught and entangled in one of the horse-hairs by which the nest was stayed and held to the limb above. There it hung bruised and dead, gibbeted to its own cradle. This nest was the theatre of another little tragedy later in the season. Some time in August a bluebird, indulging its propensity to peep and pry into holes and crevices, alighted upon it and probably inspected the interior ; but by some unlucky move it got its wings entangled in this same fatal horse-hair. Its efforts to free itself appeared only to result in its being more securely and hopelessly bound ; and there it perished ; and there its form, dried and embalmed by the summer heats, was yet hanging in September, the outspread wings and plumage showing nearly as bright as in life.

A correspondent writes me that one of his orioles got entangled in a cord while building her nest, and that, though by the aid of a ladder he reached and liberated her, she died soon afterward. He also found a " chippie " (called also " hair-bird ") suspended from a branch by a horse-hair, beneath a partly constructed nest. I heard of a cedar-bird caught and destroyed in the same way, and of two young bluebirds, around whose legs a horse-hair had become so tightly wound that the legs withered up and dropped off. The birds became fledged, and finally left the nest with the others. Such tragedies are probably quite common.

Before the advent of civilization in this country the oriole probably built a much deeper nest than it usually does at present. When now it builds in remote trees and along the borders of the woods, its nest, I have noticed, is long and gourd-shaped ; but in orchards and near dwellings it is only a deep cup or pouch. It shortens it up in proportion as the danger lessens. Probably a succession of disastrous years, like the one under review, would cause it to lengthen it again beyond the reach of owl's talons or jay-bird's beak.

The first song sparrow's nest I observed in the spring of 1881 was in a field under a fragment of a board, the board being raised from the ground a couple of inches by two poles. It had its full complement of eggs, and probably sent forth a brood of young birds, though as to this I cannot speak positively, as I neglected to observe it further. It was well sheltered and concealed, and was not easily come at by any of its natural enemies, save snakes and weasels. But concealment often avails little. In May, a song sparrow, which had evidently met with disaster earlier in the season, built its nest in a thick mass of woodbine against the side of my house, about fifteen feet from the ground. Perhaps it took the hint from its cousin, the English sparrow. The nest was admirably placed, protected from the storms by the overhanging eaves and from all eyes by the thick screen of leaves. Only by patiently watching the suspicious bird, as she lingered near with food in her beak, did I discover its whereabouts. That brood is safe, I thought, beyond doubt. But it was not: the nest was pillaged one night, either by an owl, or else by a rat that had climbed into the vine, seeking an entrance to the house. The mother bird, after reflecting upon her ill-luck about a week, seemed to resolve to try a different system of tactics, and to throw all appearances of concealment aside. She built a nest a few yards from the house, beside the drive, upon a smooth piece of greensward. There was not a weed or a shrub or anything whatever to conceal it or mark its site. The structure was completed, and incubation had begun, before I discovered what was going on. "Well, well," I said, looking down upon the bird almost at my feet, "this is going to the other extreme indeed ; now the cats will have you." The desperate little bird sat there day after day, looking like a brown leaf pressed down in the short green grass. As the weather grew hot, her position became very trying. It was no longer a question of keeping the eggs warm, but of keeping them from roasting. The sun had no mercy on her, and she fairly panted in the middle of the day. In such an emergency the male robin has been known to perch above the sitting female and shade her with his outstretched wings. But in this case there was no perch for the male bird, had he been disposed to make a sunshade of himself. I thought to lend a hand in this direction myself, and so stuck a leafy twig beside the nest. This was probably an unwise interference: it guided disaster to the spot ; the nest was broken up, and the mother bird was probably caught, as I never saw her afterward.

For several previous summers a pair of kingbirds had reared, unmolested, a brood of young in an apple-tree, only a few yards from the house; but during this season disaster overtook them also. The nest was completed, the eggs laid, and incubation had just begun, when, one morning about sunrise, I heard loud cries of distress and alarm proceed from the old apple-tree. Looking out of the window, I saw a crow, which I knew to be a fish crow, perched upon the edge of the nest, hastily bolting the eggs. The parent birds, usually so ready for the attack, seemed overcome with grief and alarm. They fluttered about in the most helpless and bewildered manner, and it was not till the robber fled on my approach that they recovered themselves and charged upon him. The crow scurried away with upturned, threatening head, the furious kingbirds fairly upon his back. The pair lingered around their desecrated nest for several days, almost silent, and saddened by their loss, and then disappeared. They probably made another trial elsewhere.

THE TRAGEDIES OF THE NESTS

The fish crow fishes only when it has destroyed all the eggs and young birds it can find. It is the most despicable thief and robber among our feathered creatures. From May to August it is gorged with the fledgelings of the nest. It is fortunate that its range is so limited. In size it is smaller than the common crow, and it is a much less noble and dignified bird. Its caw is weak and feminine, — a sort of split and abortive caw, that stamps it the sneak-thief it is. This crow is common farther south, but is not found in this State, so far as I have observed, except in the valley of the Hudson.

One season a pair of them built a nest in a Norway spruce that stood amid a dense growth of other ornamental trees near a large unoccupied house. They sat down amid plenty. The wolf established himself in the fold. The many birds — robins, thrushes, finches, vireos, pewees — that seek the vicinity of dwellings (especially of these large country residences with their many trees and park-like grounds), for the greater safety of their eggs and young, were the easy and convenient victims of these robbers. They plundered right and left, and were not disturbed till their young were nearly fledged, when some boys, who had long before marked them as their prize, rifled the nest.

The song-birds nearly all build low; their cradle is not upon the treetop. It is only birds of prey that fear danger from below more than from above,

and that seek the higher branches for their nests.
A line five feet from the ground would run above
more than half the nests, and one ten feet would
bound more than three fourths of them. It is only
the oriole, the wood pewee, the tanager, the warbling vireo, and two or three warblers, that, as a
rule, go higher than this. The crows and jays and
other enemies of the birds have learned to explore
this belt pretty thoroughly. But the leaves and
the protective coloring of most nests baffle them as
effectually, no doubt, as they do the professional
oölogist. The nest of the red-eyed vireo is one of
the most artfully placed in the wood. It is just
beyond the point where the eye naturally pauses
in its search; namely, on the extreme end of the
lowest branch of the tree, usually four or five feet
from the ground. One looks up and down and
through the tree, — shoots his eye-beams into it as
he might discharge his gun at some game hidden
there, but the drooping tip of that low horizontal
branch, — who would think of pointing his piece
just there? If a crow or other marauder were to
alight upon the branch or upon those above it, the
nest would be screened from him by the large leaf
that usually forms a canopy immediately above it.
The nest-hunter, standing at the foot of the tree
and looking straight before him, might discover it
easily, were it not for its soft, neutral gray tint
which blends so thoroughly with the trunks and

branches of trees. Indeed, I think there is no nest in the woods — no arboreal nest — so well concealed. The last one I saw was pendent from the end of a low branch of a maple, that nearly grazed the clap-boards of an unused hay-barn in a remote back-woods clearing. I peeped through a crack, and saw the old birds feed the nearly fledged young within a few inches of my face. And yet the cow-bird finds this nest and drops her parasitical egg in it. Her tactics in this as in other cases are probably to watch the movements of the parent bird. She may often be seen searching anxiously through the trees or bushes for a suitable nest, yet she may still oftener be seen perched upon some good point of observation watching the birds as they come and go about her. There is no doubt that, in many cases, the cowbird makes room for her own illegitimate egg in the nest by removing one of the bird's own. When the cowbird finds two or more eggs in a nest in which she wishes to deposit her own, she will remove one of them. I found a sparrow's nest with two sparrow's eggs and one cowbird's egg, and another egg lying a foot or so below it on the ground. I replaced the ejected egg, and the next day found it again removed, and another cowbird's egg in its place. I put it back the second time, when it was again ejected, or destroyed, for I failed to find it anywhere. Very alert and sensitive birds, like the warblers, often bury the strange egg beneath a second nest built on top of the old. A lady living in the suburbs of an eastern city, one morning heard cries of distress from a pair of house wrens that had a nest in a honeysuckle on her front porch. On looking out of the window, she beheld this little comedy, — comedy from her point of view, but no doubt grim tragedy from the point of view of the wrens: a cowbird with a wren's egg in its beak running rapidly along the walk, with the outraged wrens forming a procession behind it, screaming, scolding, and gesticulating as only these voluble little birds can. The cowbird had probably been surprised in the act of violating the nest, and the wrens were giving her a piece of their minds.

Every cowbird is reared at the expense of two or more song-birds. For every one of these dusky little pedestrians there amid the grazing cattle there are two or more sparrows, or vireos, or warblers, the less. It is a big price to pay, — two larks for a bunting, — two sovereigns for a shilling; but Nature does not hesitate occasionally to contradict herself in just this way. The young of the cowbird is disproportionately large and aggressive, one might say hoggish. When disturbed, it will clasp the nest and scream and snap its beak threateningly. One hatched out in a song sparrow's nest which was under my observation, and would soon have overridden and overborne the young sparrow which came out of the shell a few hours later, had I not interfered from time to time and lent the young sparrow a helping hand. Every day I would visit the nest and take the sparrow out from under the pot-bellied interloper, and place it on top, so that presently it was able to hold its own against its enemy. Both birds became fledged and left the nest about the same time. Whether the race was an even one after that, I know not.

I noted but two warblers' nests during that season, one of the black-throated blue-back and one of the redstart, — the latter built in an apple-tree but a few yards from a little rustic summer-house where I idle away many summer days. The lively little birds, darting and flashing about, attracted my attention for a week before I discovered their nest. They probably built it by working early in the morning, before I appeared upon the scene, as I never saw them with material in their beaks. Guessing from their movements that the nest was in a large maple that stood near by, I climbed the tree and explored it thoroughly, looking especially in the forks of the branches, as the authorities say these birds build in a fork. But no nest could I find. Indeed, how can one by searching find a bird's-nest? I overshot the mark; the nest was much nearer me, almost under my very nose, and I discovered it, not by searching, but by a casual glance of the eye, while thinking of other matters.

The bird was just settling upon it as I looked up from my book and caught her in the act. The nest was built near the end of a long, knotty, horizontal branch of an apple-tree, but effectually hidden by the grouping of the leaves; it had three eggs, one of which proved to be barren. The two young birds grew apace, and were out of the nest early in the second week; but something caught one of them the first night. The other probably grew to maturity, as it disappeared from the vicinity with its parents after some days.

The blue-back's nest was scarcely a foot from the ground, in a little bush situated in a low, dense wood of hemlock and beech and maple amid the Catskills, — a deep, massive, elaborate structure, in which the sitting bird sank till her beak and tail alone were visible above the brim. It was a misty chilly day when I chanced to find the nest, and the mother bird knew instinctively that it was not prudent to leave her four half-incubated eggs uncovered and exposed for a moment. When I sat down near the nest, she grew very uneasy, and, after trying in vain to decoy me away by suddenly dropping from the branches and dragging herself over the ground as if mortally wounded, she approached and timidly and half doubtingly covered her eggs within two yards of where I sat. I disturbed her several times, to note her ways. There came to be something almost appealing in her looks and manner, and she would keep her place on her precious eggs till my outstretched hand was within a few feet of her. Finally, I covered the cavity of the nest with a dry leaf. This she did not remove with her beak, but thrust her head deftly beneath it and shook it off upon the ground. Many of her sympathizing neighbors, attracted by her alarm note, came and had a peep at the intruder, and then flew away, but the male bird did not appear upon the scene. The final history of this nest I am unable to give, as I did not again visit it till late in the season, when, of course, it was empty.

Years pass without my finding a brown thrasher's nest; it is not a nest you are likely to stumble upon in your walk; it is hidden as a miser hides his gold, and watched as jealously. The male pours out his rich and triumphant song from the tallest tree he can find, and fairly challenges you to come and look for his treasures in his vicinity. But you will not find them if you go. The nest is somewhere on the outer circle of his song; he is never so imprudent as to take up his stand very near it. The artists who draw those cozy little pictures of a brooding mother bird, with the male perched but a yard away in full song, do not copy from nature. The thrasher's nest I found was thirty or forty rods from the point where the male was wont to indulge in his brilliant recitative. It was in an open field under a low ground-juniper. My dog disturbed the sitting bird as I was passing near. The nest could be seen only by lifting up and parting away the branches. All the arts of concealment had been carefully studied. It was the last place you would think of looking, and, if you did look, nothing was visible but the dense green circle of the low-spreading juniper. When you approached, the bird would keep her place till you had begun to stir the branches, when she would start out, and, just skimming the ground, make a bright brown line to the near fence and bushes. I confidently expected that this nest would escape molestation, but it did not. Its discovery by myself and dog probably opened the door for ill-luck, as one day, not long afterward, when I peeped in upon it, it was empty. The proud song of the male had ceased from his accustomed tree, and the pair were seen no more in that vicinity.

The phœbe-bird is a wise architect, and perhaps enjoys as great an immunity from danger, both in its person and its nest, as any other bird. Its modest, ashen-gray suit is the color of the rocks where it builds, and the moss of which it makes such free use gives to its nest the look of a natural growth or accretion. But when it comes into the barn or under the shed to build, as it so frequently does, the moss is rather out of place. Doubtless in time the bird will take the hint, and when she builds in such places will leave the moss out. I noted but two nests the summer I am speaking of: one in a barn failed of issue, on account of the rats, I suspect, though the little owl may have been the depredator ; the other, in the woods, sent forth three young. This latter nest was most charmingly and ingeniously placed. I discovered it while in quest of pond-lilies, in a long, deep, level stretch of water in the woods. A large tree had blown over at the edge of the water, and its dense mass of up-turned roots, with the black, peaty soil filling the interstices, was like the fragment of a wall several feet high, rising from the edge of the languid current. In a niche in this earthy wall, and visible and accessible only from the water, a phœbe had built her nest and reared her brood. I paddled my boat up and came alongside prepared to take the family aboard. The young, nearly ready to fly, were quite undisturbed by my presence, hav-ing probably been assured that no danger need be apprehended from that side. It was not a likely place for minks, or they would not have been so secure.

I noted but one nest of the wood pewee, and that, too, like so many other nests, failed of issue. It was saddled upon a small dry limb of a plane-tree that stood by the roadside, about forty feet from the ground. Every day for nearly a week, as I passed by, I saw the sitting bird upon the nest. Then one morning she was not in her place, and on examination the nest proved to be empty, — robbed, I had no doubt, by the red squirrels, as they were very abundant in its vicinity, and appeared to make a clean sweep of every nest. The wood pewee builds an exquisite nest, shaped and finished as if cast in a mould. It is modeled without and within with equal neatness and art, like the nest of the hummingbird and the little gray gnatcatcher. The material is much more refractory than that used by either of these birds, being, in the present case, dry, fine cedar twigs; but these were bound into a shape as rounded and compact as could be moulded out of the most plastic material. Indeed, the nest of this bird looks precisely like a large, lichen-covered, cup-shaped excrescence of the limb upon which it is placed. And the bird, while sitting, seems entirely at her ease. Most birds seem to make very hard work of incubation. It is a kind of martyrdom which appears to tax all their powers of endurance. They have such a fixed, rigid, predetermined look, pressed down into the nest and as motionless as if made of cast-iron. But the wood pewee is an exception. She is largely visible above the rim of the nest. Her attitude is easy and graceful; she moves her head this way and that, and seems to take note of whatever goes on about her; and if her neighbor were to drop in for a little social chat, she could doubtless do her part. In fact, she makes light and easy work of what, to most other birds, is such a serious and engrossing matter. If it does not look like play with her, it at least looks like leisure and quiet contemplation.

There is no nest-builder that suffers more from crows and squirrels and other enemies than the wood thrush. It builds as openly and unsuspiciously as if it thought all the world as honest as itself. Its favorite place is the fork of a sapling, eight or ten feet from the ground, where it falls an easy prey to every nest-robber that comes prowling through the woods and groves. It is not a bird that skulks and hides, like the catbird, the brown thrasher, the chat, or the chewink, and its nest is not concealed with the same art as theirs. Our thrushes are all frank, open-mannered birds; but the veery and the hermit build upon the ground, where they at least escape the crows, owls, and jays, and stand a better chance to be overlooked by the red squirrel and weasel also; while the robin seeks the protection of dwellings and outbuildings. For years I have not known the nest of a wood thrush to succeed. During the season referred to I observed but two, both apparently a second attempt, as the season was well advanced, and both failures. In one case, the nest was placed in a branch that an apple-tree, standing near a dwelling, held out over the highway. The structure was barely ten feet above the middle of the road, and would just escape a passing load of hay. It was made conspicuous by the use of a large fragment of newspaper in its foundation, — an unsafe material to build upon in most cases. Whatever else the press may guard, this particular newspaper did not guard this nest from harm. It saw the egg and probably the chick, but not the fledgeling. A murderous deed was committed above the public highway, but whether in the open day or under cover of darkness I have no means of knowing. The frisky red squirrel was doubtless the culprit. The other nest was in a maple sapling, within a few yards of the little rustic summer-house already referred to. The first attempt of the season, I suspect, had failed in a more secluded place under the hill; so the pair had come up nearer the house for protection. The male sang in the trees near by for several days before I chanced to see the nest. The very morning, I think, it was finished I saw a red squirrel exploring a tree but a few yards away; he probably knew what the singing meant as well as I did. I did not see the inside of the nest, for it was almost instantly deserted, the female having probably laid a single egg, which the squirrel had devoured.

If I were a bird, in building my nest I should follow the example of the bobolink, placing it in the midst of a broad meadow, where there was no spear of grass, or flower, or growth unlike another to mark its site. I judge that the bobolink escapes the dangers to which I have adverted as few or no

other birds do. Unless the mowers come along at an
earlier date than she has anticipated, that is, before
July 1, or a skunk goes nosing through the grass,
which is unusual, she is as safe as bird well can be
in the great open of nature. She selects the most
monotonous and uniform place she can find amid
the daisies or the timothy and clover, and places
her simple structure upon the ground in the midst
of it. There is no concealment, except as the great
conceals the little, as the desert conceals the pebble,
as the myriad conceals the unit. You may find the
nest once, if your course chances to lead you across
it, and your eye is quick enough to note the silent
brown bird as she darts swiftly away; but step
three paces in the wrong direction, and your search
will probably be fruitless. My friend and I found
a nest by accident one day, and then lost it again
one minute afterward. I moved away a few yards
to be sure of the mother bird, charging my friend
not to stir from his tracks. When I returned, he had
moved two paces, he said (he had really moved
four), and we spent a half hour stooping over the
daisies and the buttercups, looking for the lost clew.
We grew desperate, and fairly felt the ground over
with our hands, but without avail. I marked the
spot with a bush, and came the next day, and, with
the bush as a centre, moved about it in slowly
increasing circles, covering, I thought, nearly every
inch of ground with my feet, and laying hold of it

with all the visual power I could command, till my patience was exhausted, and I gave up, baffled. I began to doubt the ability of the parent birds themselves to find it, and so secreted myself and watched. After much delay, the male bird appeared with food in his beak, and, satisfying himself that the coast was clear, dropped into the grass which I had trodden down in my search. Fastening my eye upon a particular meadow-lily, I walked straight to the spot, bent down, and gazed long and intently into the grass. Finally my eye separated the nest and its young from its surroundings. My foot had barely missed them in my search, but by how much they had escaped my eye I could not tell. Probably not by distance at all, but simply by unrecognition. They were virtually invisible. The dark gray and yellowish brown dry grass and stubble of the meadow-bottom were exactly copied in the color of the half-fledged young. More than that, they hugged the nest so closely and formed such a compact mass, that though there were five of them, they preserved the unit of expression, — no single head or form was defined; they were one, and that one was without shape or color, and not separable, except by closest scrutiny, from the one of the meadow-bottom. That nest prospered, as bobolinks' nests doubtless generally do; for, notwithstanding the enormous slaughter of the birds during their fall migrations by Southern sportsmen, the bobolink appears to hold its own, and its music does not diminish in our Northern meadows.

Birds with whom the struggle for life is the sharpest seem to be more prolific than those whose nest and young are exposed to fewer dangers. The robin, the sparrow, the pewee, will rear, or make the attempt to rear, two and sometimes three broods in a season; but the bobolink, the oriole, the kingbird, the goldfinch, the cedar-bird, the birds of prey, and the woodpeckers, that build in safe retreats in the trunks of trees, have usually but a single brood. If the bobolink reared two broods, our meadows would swarm with them.

I noted three nests of the cedar-bird in August in a single orchard, all productive, but each with one or more unfruitful eggs in it. The cedar-bird is the most silent of our birds, having but a single fine note, so far as I have observed, but its manners are very expressive at times. No bird known to me is capable of expressing so much silent alarm while on the nest as this bird. As you ascend the tree and draw near it, it depresses its plumage and crest, stretches up its neck, and becomes the very picture of fear. Other birds, under like circumstances, hardly change their expression at all till they launch into the air, when by their voice they express anger rather than alarm.

I have referred to the red squirrel as a destroyer of the eggs and young of birds. I think the mis-

chief it does in this respect can hardly be overesti-
mated. Nearly all birds look upon it as their
enemy, and attack and annoy it when it appears
near their breeding haunts. Thus, I have seen the
pewee, the cuckoo, the robin, and the wood thrush
pursuing it with angry voice and gestures. A friend
of mine saw a pair of robins attack one in the top
of a tall tree so vigorously that they caused it to
lose its hold, when it fell to the ground, and was so
stunned by the blow as to allow him to pick it up.
If you wish the birds to breed and thrive in your
orchards and groves, kill every red squirrel that infests the place; kill every weasel also. The weasel
is a subtle and arch enemy of the birds. It climbs
trees and explores them with great ease and nimble-
ness. I have seen it do so on several occasions.
One day my attention was arrested by the angry
notes of a pair of brown thrashers that were flitting
from bush to bush along an old stone row in a remote field. Presently I saw what it was that excited
them, — three large red weasels, or ermines, coming
along the stone wall, and leisurely and half playfully exploring every tree that stood near it. They
had probably robbed the thrashers. They would
go up the trees with great ease, and glide serpent-
like out upon the main branches. When they de-
scended the tree, they were unable to come straight
down, like a squirrel, but went around it spirally.
How boldly they thrust their heads out of the wall.

and eyed me and sniffed me as I drew near, — their round, thin ears, their prominent, glistening, bead-like eyes, and the curving, snake-like motions of the head and neck being very noticeable. They looked like blood-suckers and egg-suckers. They suggested something extremely remorseless and cruel. One could understand the alarm of the rats when they discover one of these fearless, subtle, and circumventing creatures threading their holes. To flee must be like trying to escape death itself. I was one day standing in the woods upon a flat stone, in what at certain seasons was the bed of a stream, when one of these weasels came undulat-ing along and ran under the stone upon which I was standing. As I remained motionless, he thrust out his wedge-shaped head, and turned it back above the stone as if half in mind to seize my foot; then he drew back, and presently went his way. These weasels often hunt in packs like the Brit-ish stoat. When I was a boy, my father one day armed me with an old musket and sent me to shoot chipmunks around the corn. While watching the squirrels, a troop of weasels tried to cross a bar-way where I sat, and were so bent on doing it that I fired at them, boy-like, simply to thwart their purpose. One of the weasels was disabled by my shot, but the troop were not discouraged, and, after making several feints to cross, one of them seized the wounded one and bore it over, and the pack disappeared in the wall on the other side.

Let me conclude this chapter with two or three more notes about this alert enemy of the birds and the lesser animals, the weasel.

A farmer one day heard a queer growling sound in the grass ; on approaching the spot he saw two weasels contending over a mouse ; both had hold of the mouse, pulling in opposite directions, and they were so absorbed in the struggle that the farmer cautiously put his hands down and grabbed them both by the back of the neck. He put them in a cage, and offered them bread and other food. This they refused to eat, but in a few days one of them had eaten the other up, picking his bones clean, and leaving nothing but the skeleton.

The same farmer was one day in his cellar when two rats came out of a hole near him in great haste, and ran up the cellar wall and along its top till they came to a floor timber that stopped their progress, when they turned at bay, and looked excitedly back along the course they had come. In a moment a weasel, evidently in hot pursuit of them, came out of the hole, and, seeing the farmer, checked his course and darted back. The rats had doubtless turned to give him fight, and would probably have been a match for him.

The weasel seems to track its game by scent. A hunter of my acquaintance was one day sitting in the woods, when he saw a red squirrel run with great speed up a tree near him, and out upon a long branch, from which he leaped to some rocks, and disappeared beneath them. In a moment a weasel came in full course upon his trail, ran up the tree, then out along the branch, from the end of which he leaped to the rocks as the squirrel did, and plunged beneath them.

Doubtless the squirrel fell a prey to him. The squirrel's best game would have been to have kept to the higher treetops, where he could easily have distanced the weasel. But beneath the rocks he stood a very poor chance. I have often wondered what keeps such an animal as the weasel in check, for weasels are quite rare. They never need go hungry, for rats and squirrels and mice and birds are everywhere. They probably do not fall a prey to any other animal, and very rarely to man. But the circumstances or agencies that check the increase of any species of animal or bird are, as Darwin says, very obscure and but little known.

A RIVER VIEW

*In the gazebo at Riverby
on the west bank of the Hudson River*

EDITOR'S NOTE

"A River View" was published in *Scribner's* in 1880 and reprinted in *Signs and Seasons* in 1886. Burroughs had been observing the Hudson River from his home on its west bank for thirteen years when he completed the final version of this essay for book publication. He had begun keeping a journal on May 13, 1876, three years after moving to Riverby, and its pages provided him with numerous detailed descriptions of seasonal changes and events for use in the essay. His distinction as a writer becomes evident in the use he makes of these details. While remaining vividly particular, the Hudson becomes a metaphor for nature itself as Burroughs transforms it into a thing of life, a mass of clouds, a strip of firmament dotted with stars and moons, an eruption from a frost volcano, the walls of the celestial city, a phantom thunderstorm, a field of icy wheat and corn, a crystal plain of frost ferns, and a wild, savage arm of the sea. Burroughs draws his concluding observations of the river from the perspective of geology, a science that increasingly inter-

EDITOR'S NOTE

ested him as he grew older. In this essay his scientific perspective and Ovidian vision merge into an effective evocation of the moods and metamorphoses of a great river.

F.B.

A RIVER VIEW

A SMALL river or stream flowing by one's door has many attractions over a large body of water like the Hudson. One can make a companion of it ; he can walk with it and sit with it, or lounge on its banks, and feel that it is all his own. It becomes something private and special to him. You cannot have the same kind of attachment and sympathy with a great river; it does not flow through your affections like a lesser stream. The Hudson is a long arm of the sea, and it has something of the sea's austerity and grandeur. I think one might spend a lifetime upon its banks without feeling any sense of ownership in it, or becoming at all intimate with it: it keeps one at arm's length. It is a great highway of travel and of commerce ; ships from all parts of our seaboard plow its waters.

But there is one thing a large river does for one that is beyond the scope of the companionable streams, — it idealizes the landscape, it multiplies and heightens the beauty of the day and of the season. A fair day it makes more fair, and a wild and tempestuous day it makes more wild and tem-

pestuous. It takes on so quickly and completely
the mood and temper of the sky above. The storm
is mirrored in it, and the wind chafes it into foam.
The face of winter it makes doubly rigid and corpse-
like. How stark and still and white it lies there!
But of a bright day in spring, what life and light
possess it! How it enhances or emphasizes the
beauty of those calm, motionless days of summer
or fall, — the broad, glassy surface perfectly dupli-
cating the opposite shore, sometimes so smooth that
the finer floating matter here and there looks like
dust upon a mirror ; the becalmed sails standing
this way and that, drifting with the tide. Indeed,
nothing points a calm day like a great motionless
sail ; it is such a conspicuous bid for the breeze
which comes not.

I have observed that when the river is roily, the
fact is not noticeable on a calm day; a glassy surface is a kind of mask. But when the breeze comes
and agitates it a little, its real color comes out.

" Immortal water," says Thoreau, " alive to the
superficies." How sensitive and tremulous and
palpitating this great river is! It is only in certain lights, on certain days, that we can see how it
quivers and throbs. Sometimes you can see the
subtle tremor or impulse that travels in advance of
the coming steamer and prophesies of its coming.
Sometimes the coming of the flood-tide is heralded
in the same way. Always, when the surface is

calm enough and the light is favorable, the river seems shot through and through with tremblings and premonitions.

The river never seems so much a thing of life as in the spring when it first slips off its icy fetters. The dead comes to life before one's very eyes. The rigid, pallid river is resurrected in a twinkling. You look out of your window one moment, and there is that great, white, motionless expanse; you look again, and there in its place is the tender, dimpling, sparkling water. But if your eyes are sharp, you may have noticed the signs all the forenoon; the time was ripe, the river stirred a little in its icy shroud, put forth a little streak or filament of blue water near shore, made breathing-holes. Then, after a while, the ice was rent in places, and the edges crushed together or shoved one slightly upon the other ; there was apparently something growing more and more alive and restless underneath. Then suddenly the whole mass of the ice from shore to shore begins to move downstream,— very gently, almost imperceptibly at first, then with a steady, deliberate pace that soon lays bare a large expanse of bright, dancing water. The island above keeps back the northern ice, and the ebb tide makes a clean sweep from that point south for a few miles, until the return of the flood, when the ice comes back.

After the ice is once in motion, a few hours suffice to break it up pretty thoroughly. Then what a wild, chaotic scene the river presents: in one part of the day the great masses hurrying downstream, crowding and jostling each other, and struggling for the right of way; in the other, all running upstream again, as if sure of escape in that direction. Thus they race up and down, the sport of the ebb and flow; but the ebb wins each time by some distance. Large fields from above, where the men were at work but a day or two since, come down; there is their pond yet clearly defined and full of marked ice; yonder is a section of their canal partly filled with the square blocks on their way to the elevators; a piece of a race-course, or a part of a road where teams crossed, comes drifting by. The people up above have written their winter pleasure and occupations upon this page, and we read the signs as the tide bears it slowly past. Some calm, bright days the scattered and diminished masses glide by like white clouds across an April sky.

At other times, when the water is black and still, the river looks like a strip of the firmament at night, dotted with stars and moons in the shape of little and big fragments of ice. One day, I remember, there came gliding into my vision a great irregular hemisphere of ice, that vividly suggested the half moon under the telescope; its white uneven surface, pitted and cracked, the jagged inner line, the outward curve, but little broken, and the blue-

. black surface upon which it lay, all recalled the
scenery of the midnight skies. It is only in exceptionally calm weather that the ice collects in these
vast masses, leaving broad expanses of water perfectly clear. Sometimes, during such weather, it
drifts by in forms that suggest the great continents,
as they appear upon the map, surrounded by the
oceans, all their capes and peninsulas, and isth-
muses and gulfs, and inland lakes and seas, vividly
reproduced.

If the opening of the river is gentle, the closing
of it is sometimes attended by scenes exactly the
reverse.

A cold wave one December was accompanied by
a violent wind, which blew for two days and two
nights. The ice formed rapidly in the river, but the
wind and waves kept it from uniting and massing.
On the second day the scene was indescribably
wild and forbidding; the frost and fury of December were never more vividly pictured: vast crumpled, spumy ice-fields interspersed with stretches
of wildly agitated water, the heaving waves thick
with forming crystals, the shores piled with frozen
foam and pulverized floes. After the cold wave had
spent itself and the masses had become united and
stationary, the scene was scarcely less wild. I fancied the plain looked more like a field of lava and
scoria than like a field of ice, an eruption from
some huge frost volcano of the north. Or did it

suggest that a battle had been fought there, and that this wild confusion was the ruin wrought by the contending forces?

No sooner has the river pulled his icy coverlid over him than he begins to snore in his winter sleep. It is a singular sound. Thoreau calls it a "whoop," Emerson a "cannonade," and in "Merlin" speaks of

> "the gasp and moan
> Of the ice-imprisoned flood."

Sometimes it is a well-defined grunt, — *e-h-h, e-h-h,* as if some ice-god turned uneasily in his bed.

One fancies the sound is like this, when he hears it in the still winter nights seated by his fireside, or else when snugly wrapped in his own bed.

One winter the river shut up in a single night, beneath a cold wave of great severity and extent. Zero weather continued nearly a week, with a clear sky and calm, motionless air; and the effect of the brilliant sun by day and of the naked skies by night upon this vast area of new black ice, one expanding it, the other contracting, was very marked.

A cannonade indeed! As the morning advanced, out of the sunshine came peal upon peal of soft mimic thunder; occasionally becoming a regular crash, as if all the ice batteries were discharged at once. As noon approached, the sound grew to one continuous mellow roar, which lessened and became more intermittent as the day waned, until about sundown it was nearly hushed. Then, as the chill of night came on, the conditions were reversed, and the ice began to thunder under the effects of contraction; cracks opened from shore to shore, and grew to be two or three inches broad under the shrinkage of the ice. On the morrow the expansion of the ice often found vent in one of these cracks, the two edges would first crush together, and then gradually overlap each other for two feet or more.

This expansive force of the sun upon the ice is sometimes enormous. I have seen the ice explode with a loud noise and a great commotion in the water, and a huge crack shoot like a thunderbolt from shore to shore, with its edges overlapping and shivered into fragments.

When unprotected by a covering of snow, the ice, under the expansive force of the sun, breaks regularly, every two or three miles, from shore to shore. The break appears as a slight ridge, formed by the edges of the overlapping ice.

This icy uproar is like thunder, because it seems to proceed from something in swift motion; you cannot locate it; it is everywhere and yet nowhere. There is something strange and phantom-like about it. To the eye all is still and rigid, but to the ear all is in swift motion.

This crystal cloud does not open and let the bolt leap forth, but walk upon it and you see the ice shot through and through in every direction with shining, iridescent lines where the force passed. These lines are not cracks which come to the surface, but spiral paths through the ice, as if the force that made them went with a twist like a rifle bullet. In places several of them run together, when they make a track as broad as one's hand.

Sometimes, when I am walking upon the ice and this sound flashes by me, I fancy it is like the stroke of a gigantic skater, one who covers a mile at a stride and makes the crystal floor ring beneath him. I hear his long tapering stroke ring out just beside me, and then in a twinkling it is half a mile away.

A fall of snow, and this icy uproar is instantly hushed, the river sleeps in peace. The snow is like a coverlid, which protects the ice from the changes of temperature of the air, and brings repose to its uneasy spirit.

A dweller upon its banks, I am an interested spectator of the spring and winter harvests which its waters yield. In the stern winter nights, it is a pleasant thought that a harvest is growing down there on those desolate plains which will bring work to many needy hands by and by, and health and comfort to the great cities some months later. When the nights are coldest, the ice grows as fast as corn in July. It is a crop that usually takes two or three weeks to grow, and, if the water is very roily or brackish, even longer. Men go out from time to time and examine it, as the farmer goes out and examines his grain or grass, to see when it will do to cut. If there comes a deep fall of snow before the ice has attained much thickness, it is "pricked," so as to let the water up through and form snow-ice. A band of fifteen or twenty men, about a yard apart, each armed with a chisel-bar and marching in line, puncture the ice at each step with a single sharp thrust. To and fro they go, leaving a belt behind them that presently becomes saturated with water. But ice, to be first quality, must grow from beneath, not from above. It is a crop quite as uncertain as any other. A good yield every two or three years, as they say of wheat out West, is about all that can be counted on. When there is an abundant harvest, after the ice-houses are filled, they stack great quantities of it, as the farmer stacks his surplus hay.

The cutting and gathering of the ice enlivens these broad, white, desolate fields amazingly. One looks down upon the busy scene as from a hill-top upon a river meadow in haying time, only here the figures stand out much more sharply than they do from a summer meadow. There is the broad, straight, blue-black canal emerging into view, and running nearly across the river; this is the highway that lays open the farm. On either side lie the fields or ice-meadows, each marked out by cedar or hemlock boughs. The farther one is cut first, and, when cleared, shows a large, long, black parallelogram in the midst of the plain of snow. Then the next one is cut, leaving a strip or tongue of ice between the two for the horses to move and turn upon. Sometimes nearly two hundred men and boys, with numerous horses, are at work at once, marking, plowing, planing, scraping, sawing, hauling, chiseling; some floating down the pond on great square islands towed by a horse, or their fellow-workmen; others distributed along the canal, bending to their ice-hooks; others upon the bridges, separating the blocks with their chisel-bars; others feeding the elevators; while knots and straggling lines of idlers here and there look on in cold discontent, unable to get a job.

The best crop of ice is an early crop. Late in the season, or after January, the ice is apt to get " sunstruck," when it becomes " shaky," like a piece of poor timber. The sun, when he sets about destroying the ice, does not simply melt it from the surface, — that were a slow process; but he sends his shafts into it and separates it into spikes and needles, — in short, makes kindling-wood of it; so as to consume it the quicker.

One of the prettiest sights about the ice-harvesting is the elevator in operation. When all works well, there is an unbroken procession of the great crystal blocks slowly ascending this incline. They go up in couples, arm in arm, as it were, like friends up a stairway, glowing and changing in the sun, and recalling the precious stones that adorned the walls of the celestial city. When they reach the platform where they leave the elevator, they seem to step off like things of life and volition; they are still in pairs, and separate only as they enter upon the " runs." But here they have an ordeal to pass through, for they are subjected to a rapid inspection by a man with a sharp eye in his head and a sharp ice-hook in his hand, and the black sheep are separated from the flock; every square with a trace of sediment or earth-stain in it, whose texture is not the perfect and unclouded crystal, is rejected, and sent hurling down into the abyss. Those that pass the examination glide into the building along the gentle incline, and are switched off here and there upon branch runs, and distributed to all parts of the immense interior. When the momentum becomes too great, the blocks run over a board full of nails or spikes, that scratch their bottoms and retard their progress, giving the looker-on an uncomfortable feeling.

A beautiful phenomenon may at times be witnessed on the river in the morning after a night of extreme cold. The new black ice is found to be covered with a sudden growth of frost ferns, — exquisite fern-like formations from a half inch to an inch in length, standing singly and in clusters, and under the morning sun presenting a most novel appearance. They impede the skate, and are presently broken down and blown about by the wind.

The scenes and doings of summer are counterfeited in other particulars upon these crystal plains. Some bright, breezy day you casually glance down the river and behold a sail, — a sail like that of a pleasure yacht of summer. Is the river open again below there? is your first half-defined inquiry. But with what unwonted speed the sail is moving across the view! Before you have fairly drawn another breath it has turned, unperceived, and is shooting with equal swiftness in the opposite direction. Who ever saw such a lively sail! It does not bend before the breeze, but darts to and fro as if it moved in a vacuum, or like a shadow over a screen. Then you remember the ice-boats, and you open your eyes to the fact. Another and another come into view around the elbow, turning and flashing in the sun, and hurtling across each other's path like white-winged gulls. They turn so quickly, and dash off again at such speed, that they produce the illusion of something singularly light and intangible. In fact, an ice-boat is a sort of disembodied yacht; it is a sail on skates. The only semblance to a boat is the sail and the rudder. The platform under which the skates or runners — three in number — are rigged is broad and low; upon this the pleasure-seekers, wrapped in their furs or blankets, lie at full length, and, looking under the sail, skim the frozen surface with their eyes. The speed attained is sometimes very great, — more than a mile per minute, and sufficient to carry them ahead of the fastest express train. When going at this rate the boat will leap like a greyhound, and thrilling stories are told of the fearful crevasses, or open places in the ice, that are cleared at a bound. And yet withal she can be brought up to the wind so suddenly as to shoot the unwary occupants off, and send them skating on their noses some yards.

Navigation on the Hudson stops about the last of November. There is usually more or less floating ice by that time, and the river may close very abruptly. Beside that, new ice an inch or two thick is the most dangerous of all ; it will cut through a vessel's hull like a knife. In 1875 there was a sudden fall of the mercury the 28th of November. The hard and merciless cold came down upon the naked earth with great intensity. On the 29th the ground was a rock, and, after the sun went down, the sky all around the horizon looked like a wall of chilled iron. The river was quickly covered with great floating fields of smooth, thin ice. About three o'clock the next morning — the mercury two degrees below zero — the silence of our part of the river was suddenly broken by the alarm bell of a passing steamer; she was in the jaws of the icy legions, and was crying for help; many sleepers alongshore remembered next day that the sound of a bell had floated across their dreams, without arousing them. One man was awakened before long by a loud pounding at his door. On opening it, a tall form, wet and icy, fell in upon him with the cry, "The Sunnyside is sunk!" The man proved to be one of her officers, and was in quest of help. He had made his way up a long hill through the darkness, his wet clothes freezing upon him, and his strength gave way the moment succor was found. Other dwellers in the vicinity were aroused, and with their boats rendered all the assistance possible. The steamer sank but a few yards from shore, only a part of her upper deck remaining above water, yet a panic among the passengers — the men behaving very badly — swamped the boats as they were being filled with the women, and a dozen or more persons were drowned.

When the river is at its wildest, usually in March, the eagles appear. They prowl about amid the ice-floes, alighting upon them or flying heavily above them in quest of fish, or a wounded duck or other game.

I have counted ten of these noble birds at one time, some seated grim and motionless upon cakes of ice, — usually surrounded by crows, — others flapping along, sharply scrutinizing the surface beneath. Where the eagles are, there the crows do congregate. The crow follows the eagle, as the jackal follows the lion, in hope of getting the leav-

ings of the royal table. Then I suspect the crow is
a real hero-worshiper. I have seen a dozen or more
of them sitting in a circle about an eagle upon the
ice, all with their faces turned toward him, and
apparently in silent admiration of the dusky king.

The eagle seldom or never turns his back upon
a storm. I think he loves to face the wildest elemental commotion. I shall long carry the picture
of one I saw floating northward on a large raft of
ice one day, in the face of a furious gale of snow.
He stood with his talons buried in the ice, his head
straight out before him, his closed wings showing
their strong elbows, — a type of stern defiance and
power.

This great metropolitan river, as it were, with
its floating palaces, and shores lined with villas, is
thus an inlet and a highway of the wild and the
savage. The wild ducks and geese still follow it
north in spring, and south in the fall. The loon
pauses in his migrations and disports himself in its
waters. Seals and otters are occasionally seen in it.

Of the Hudson it may be said that it is a very
large river for its size, — that is, for the quantity
of water it discharges into the sea. Its water-shed
is comparatively small, — less, I think, than that of
the Connecticut.

It is a huge trough with a very slight incline,
through which the current moves very slowly, and
which would fill from the sea were its supplies from

the mountains cut off. Its fall from Albany to the bay is only about five feet. Any object upon it, drifting with the current, progresses southward no more than eight miles in twenty-four hours. The ebb tide will carry it about twelve miles, and the flood set it back from seven to nine. A drop of water at Albany, therefore, will be nearly three weeks in reaching New York, though it will get pretty well pickled some days earlier.

Some rivers by their volume and impetuosity penetrate the sea, but here the sea is the aggressor, and sometimes meets the mountain water nearly halfway.

This fact was illustrated a few years ago, when the basin of the Hudson was visited by one of the most severe droughts ever known in this part of the State. In the early winter, after the river was frozen over above Poughkeepsie, it was discovered that immense numbers of fish were retreating up-stream before the slow encroachment of the salt water. There was a general exodus of the finny tribes from the whole lower part of the river ; it was like the spring and fall migration of the birds, or the fleeing of the population of a district before some approaching danger ; vast swarms of catfish, white and yellow perch, and striped bass were *en route* for the fresh water farther north. When the people alongshore made the discovery, they turned out as they do in the rural districts when the pigeons appear, and, with small gillnets let down through holes in the ice, captured them in fabulous numbers. On the heels of the retreating perch and catfish came the denizens of salt water, and codfish were taken ninety miles above New York. When the February thaw came, and brought up the volume of fresh water again, the sea brine was beaten back, and the fish, what were left of them, resumed their old feeding-grounds.

It is this character of the Hudson, this encroachment of the sea upon it, that has led Professor Newberry to speak of it as a drowned river. We have heard of drowned lands, but here is a river overflowed and submerged in the same manner. It is quite certain, however, that this has not always been the character of the Hudson. Its great trough bears evidence of having been worn to its present dimensions by much swifter and stronger currents than those that course through it now. Hence Professor Newberry has advanced the bold and striking theory that in pre-glacial times this part of the continent was several hundred feet higher than at present, and that the Hudson was then a very large and rapid stream, that drew its main supplies from the basin of the Great Lakes through an ancient river-bed that followed pretty nearly the line of the present Mohawk; in other words, that the waters of the St. Lawrence once found an outlet through this channel, debouching into the ocean from a broad, littoral plain, at a point eighty miles southeast of New York, where the sea now rolls five hundred feet deep. According to the soundings of the coast survey, this ancient bed of the Hudson is distinctly marked upon the ocean floor to the point indicated.

To the gradual subsidence of this part of the continent, in connection with the great changes wrought by the huge glacier that crept down from the north during what is called the ice period, is owing the character and aspects of the Hudson as we see and know them. The Mohawk valley was filled up by the drift, and the pent-up waters of the Great Lakes found an opening through what is now the St. Lawrence. The trough of the Hudson was also partially filled, and has remained so to the present day. There is, perhaps, no point in the river where the mud and clay are not from two to three times as deep as the water.

That ancient and grander Hudson lies back of us several hundred thousand years, — perhaps more, for a million years are but as one tick of the time-piece of the Lord; yet even *it* was a juvenile compared with some of the rocks and mountains the Hudson of to-day mirrors. The Highlands date from the earliest geological age, — the primary; the river — the old river — from the latest, the tertiary; and what that difference means in terrestrial years hath not entered into the mind of man to conceive. Yet how the venerable mountains open their ranks for the stripling to pass through. Of course the river did not force its way through this barrier, but has doubtless found an opening there of which it has availeα itself, and which it has enlarged.

In thinking of these things, one only has to allow time enough, and the most stupendous changes in the topography of the country are as easy and natural as the going out or the coming in of spring or summer. According to the authority above referred to, that part of our coast that flanks the mouth of the Hudson is still sinking at the rate of a few inches per century, so that in the twinkling of a hundred thousand years or so the sea will completely submerge the city of New York, the top of Trinity Church steeple alone standing above the flood. We who live so far inland, and sigh for the salt water, need only to have a little patience, and we shall wake up some fine morning and find the surf beating upon our doorsteps.

THE HEART OF
THE SOUTHERN CATSKILLS

Cooking his favorite "brigand steak"

EDITOR'S NOTE

In late spring 1885 Burroughs wrote to his friend Myron Benton, "I feel budding within me an expedition to Slide mountain, and to some of those trout streams. How do you feel?" Myron Benton, who had written Burroughs his first fan letter and remained his friend for fifty years, joined William and John Van Benschoten for the expedition with Burroughs on June 7, 1885. It was Burroughs's first ascent of Slide Mountain, the highest peak in the Catskills. He describes how he and his companions, like many mountain climbers, experience the pleasures and discomforts of seemingly moving back in time and north in latitude while gaining altitude. First they overtake spring, then find themselves in a November storm, before descending back to a valley with its summer trout and wild strawberries. Always interested in the particular and the local, Burroughs describes what he wishes might be called the Slide Mountain thrush, a gray-cheeked thrush whose song he renders with the typical attentiveness that earned him his reputation as a master of bird song. Burroughs is interested not only in the bird

EDITOR'S NOTE

and its song but also in the relationship of both to a particular environment. The essay also contains rare instances of Burroughs's explicitly reflecting on the relationship between civilization and wilderness. After this ascent, Burroughs climbed Slide Mountain many times, and near where he camped on the peak, a bronze tablet now honors him for introducing the mountain to the world. Slide Mountain is now part of the Burroughs Range in the Catskills. This essay first appeared in *Century* in August 1888 and was reprinted in the volume *Riverby* in 1894.

F.B.

THE HEART OF THE SOUTHERN CATSKILLS

ON looking at the southern and more distant Catskills from the Hudson River on the east, or on looking at them from the west from some point of vantage in Delaware County, you see, amid the group of mountains, one that looks like the back and shoulders of a gigantic horse. The horse has got his head down grazing; the shoulders are high, and the descent from them down his neck very steep; if he were to lift up his head, one sees that it would be carried far above all other peaks, and that the noble beast might gaze straight to his peers in the Adirondacks or the White Mountains. But the lowered head never comes up; some spell or enchantment keeps it down there amid the mighty herd; and the high round shoulders and the smooth strong back of the steed are alone visible. The peak to which I refer is Slide Mountain, the highest of the Catskills by some two hundred feet, and probably the most inaccessible; certainly the hardest to get a view of, it is hedged about so completely by other peaks, — the greatest mountain of them all, and apparently the least willing to be seen; only at a distance of thirty or forty miles is it seen to stand up above all other peaks. It takes its name from a landslide which occurred many years ago down its steep northern side, or down the neck of the grazing steed. The mane of spruce and balsam fir was stripped away for many hundred feet, leaving a long gray streak visible from afar.

Slide Mountain is the centre and the chief of the southern Catskills. Streams flow from its base, and from the base of its subordinates, to all points of the compass, — the Rondout and the Neversink to the south ; the Beaverkill to the west ; the Esopus to the north ; and several lesser streams to the east. With its summit as the centre, a radius of ten miles would include within the circle described but very little cultivated land ; only a few poor, wild farms in some of the numerous valleys. The soil is poor, a mixture of gravel and clay, and is subject to slides. It lies in the valleys in ridges and small hillocks, as if dumped there from a huge cart. The tops of the southern Catskills are all capped with a kind of conglomerate, or "pudden stone," — a rock of cemented quartz pebbles which underlies the coal measures. This rock disintegrates under the action of the elements, and the sand and gravel which result are carried into the valleys and make up the most of the soil. From the northern Catskills, so far as I know them, this rock has been swept clean.

THE SOUTHERN CATSKILLS

Low down in the valleys the old red sandstone crops out, and, as you go west into Delaware County, in many places it alone remains and makes up most of the soil, all the superincumbent rock having been carried away.

Slide Mountain had been a summons and a challenge to me for many years. I had fished every stream that it nourished, and had camped in the wilderness on all sides of it, and whenever I had caught a glimpse of its summit I had promised myself to set foot there before another season should pass. But the seasons came and went, and my feet got no nimbler, and Slide Mountain no lower, until finally, one July, seconded by an energetic friend, we thought to bring Slide to terms by approaching him through the mountains on the east. With a farmer's son for guide we struck in by way of Weaver Hollow, and, after a long and desperate climb, contented ourselves with the Wittenberg, instead of Slide. The view from the Wittenberg is in many respects more striking, as you are perched immediately above a broader and more distant sweep of country, and are only about two hundred feet lower. You are here on the eastern brink of the southern Catskills, and the earth falls away at your feet and curves down through an immense stretch of forest till it joins the plain of Shokan, and thence sweeps away to the Hudson and beyond. Slide is southwest of you, six or seven miles distant, but is visible only when you climb into a treetop. I climbed and saluted him, and promised to call next time.

We passed the night on the Wittenberg, sleeping on the moss, between two decayed logs, with balsam boughs thrust into the ground and meeting and forming a canopy over us. In coming off the mountain in the morning we ran upon a huge porcupine, and I learned for the first time that the tail of a porcupine goes with a spring like a trap. It seems to be a set-lock; and you no sooner touch with the weight of a hair one of the quills than the tail leaps up in a most surprising manner, and the laugh is not on your side. The beast cantered along the path in my front, and I threw myself upon him, shielded by my roll of blankets. He submitted quietly to the indignity, and lay very still under my blankets, with his broad tail pressed close to the ground. This I proceeded to investigate, but had not fairly made a beginning when it went off like a trap, and my hand and wrist were full of quills. This caused me to let up on the creature, when it lumbered away till it tumbled down a precipice. The quills were quickly removed from my hand, when we gave chase. When we came up to him, he had wedged himself in between the rocks so that he presented only a back bristling with quills, with the tail lying in ambush below. He had chosen his position well, and seemed to defy us. After amusing ourselves by repeatedly springing his tail and receiving the quills in a rotten stick, we made a slip-noose out of a spruce root, and, after much manœuvring, got it over his head and led him forth. In what a peevish, injured tone the creature did complain of our unfair tactics! He protested and protested, and whimpered and scolded like some infirm old man tormented by boys. His game after we led him forth was to keep himself as much as possible in the shape of a ball, but with two sticks and the cord we finally threw him over on his back and exposed his quill-less and vulnerable under side, when he fairly surrendered and seemed to say, "Now you may do with me as you like." His great chisel-like teeth, which are quite as formidable as those of the woodchuck, he does not appear to use at all in his defense, but relies entirely upon his quills, and when those fail him, he is done for.

After amusing ourselves with him awhile longer, we released him and went on our way. The trail to which we had committed ourselves led us down into Woodland Valley, a retreat which so took my eye by its fine trout brook, its superb mountain scenery, and its sweet seclusion, that I marked it for my own, and promised myself a return to it at no distant day. This promise I kept, and pitched my tent there twice during that season. Both occasions were a sort of laying siege to Slide, but we only skirmished with him at a distance; the actual assault was not undertaken. But the following year, reinforced by two other brave climbers, we determined upon the assault, and upon making it from this the most difficult side. The regular way is by Big Ingin Valley, where the climb is comparatively easy, and where it is often made by women. But from Woodland Valley only men may essay the ascent. Larkins is the upper inhabitant, and from our camping-ground near his clearing we set out early one June morning.

One would think nothing could be easier to find than a big mountain, especially when one is encamped upon a stream which he knows springs out of its very loins. But for some reason or other we had got an idea that Slide Mountain was a very slippery customer and must be approached cautiously. We had tried from several points in the valley to get a view of it, but were not quite sure we had seen its very head. When on the Wittenberg, a neighboring peak, the year before, I had caught a brief glimpse of it only by climbing a dead tree and craning up for a moment from its topmost branch. It would seem as if the mountain had taken every precaution to shut itself off from a near view. It was a shy mountain, and we were about to stalk it through six or seven miles of primitive woods, and we seemed to have some unreasonable fear that it might elude us. We had been told of parties who had essayed the ascent from this side, and had returned baffled and bewildered. In a tangle of prim-

itive woods, the very bigness of the mountain baffles
one. It is all mountain; whichever way you turn
— and one turns sometimes in such cases before he
knows it — the foot finds a steep and rugged ascent.

The eye is of little service ; one must be sure of
his bearings and push boldly on and up. One is
not unlike a flea upon a great shaggy beast, looking
for the animal's head; or even like a much smaller
and much less nimble creature, — he may waste his
time and steps, and think he has reached the head
when he is only upon the rump. Hence I questioned our host, who had several times made the
ascent, closely. Larkins laid his old felt hat upon
the table, and, placing one hand upon one side of it
and the other upon the other, said: " There Slide
lies, between the two forks of the stream, just as my
hat lies between my two hands. David will go with
you to the forks, and then you will push right on up."
But Larkins was not right, though he had traversed
all those mountains many times over. The peak we
were about to set out for did not lie between the
forks, but exactly at the head of one of them; the
beginnings of the stream are in the very path of the
slide, as we afterward found. We broke camp early
in the morning, and with our blankets strapped to
our backs and rations in our pockets for two days,
set out along an ancient and in places an obliterated
bark road that followed and crossed and recrossed
the stream. The morning was bright and warm, but

the wind was fitful and petulant, and I predicted rain. What a forest solitude our obstructed and dilapidated wood-road led us through! five miles of primitive woods before we came to the forks, three miles before we came to the "burnt shanty," a name merely, — no shanty there now for twenty-five years past. The ravages of the barkpeelers were still visible, now in a space thickly strewn with the soft and decayed trunks of hemlock-trees, and overgrown with wild cherry, then in huge mossy logs scattered through the beech and maple woods. Some of these logs were so soft and mossy that one could sit or recline upon them as upon a sofa.

But the prettiest thing was the stream soliloquizing in such musical tones there amid the moss-covered rocks and boulders. How clean it looked, what purity! Civilization corrupts the streams as it corrupts the Indian; only in such remote woods can you now see a brook in all its original freshness and beauty. Only the sea and the mountain forest brook are pure; all between is contaminated more or less by the work of man. An ideal trout brook was this, now hurrying, now loitering, now deepening around a great boulder, now gliding evenly over a pavement of green-gray stone and pebbles; no sediment or stain of any kind, but white and sparkling as snow-water, and nearly as cool. Indeed, the water of all this Catskill region is the best in the world. For the first few days, one feels as if he could almost live on the water alone; he cannot drink enough of it. In this particular it is indeed the good Bible land, "a land of brooks of water, of fountains and depths that spring out of valleys and hills."

Near the forks we caught, or thought we caught, through an opening, a glimpse of Slide. Was it Slide? was it the head, or the rump, or the shoulder of the shaggy monster we were in quest of? At the forks there was a bewildering maze of underbrush and great trees, and the way did not seem at all certain; nor was David, who was then at the end of his reckoning, able to reassure us. But in assaulting a mountain, as in assaulting a fort, boldness is the watchword. We pressed forward, following a line of blazed trees for nearly a mile, then, turning to the left, began the ascent of the mountain. It was steep, hard climbing. We saw numerous marks of both bears and deer; but no birds, save at long intervals the winter wren flitting here and there, and darting under logs and rubbish like a mouse. Occasionally its gushing, lyrical song would break the silence. After we had climbed an hour or two, the clouds began to gather, and presently the rain began to come down. This was discouraging; but we put our backs up against trees and rocks, and waited for the shower to pass.

"They were wet with the showers of the mountain, and embraced the rocks for want of shelter," as they did in Job's time. But the shower was

light and brief, and we were soon under way again.
Three hours from the forks brought us out on the
broad level back of the mountain upon which Slide,
considered as an isolated peak, is reared. After a
time we entered a dense growth of spruce which
covered a slight depression in the table of the mountain. The moss was deep, the ground spongy, the
light dim, the air hushed. The transition from the
open, leafy woods to this dim, silent, weird grove
was very marked. It was like the passage from the
street into the temple. Here we paused awhile and
ate our lunch, and refreshed ourselves with water
gathered from a little well sunk in the moss.

The quiet and repose of this spruce grove proved
to be the calm that goes before the storm. As we
passed out of it, we came plump upon the almost
perpendicular battlements of Slide. The mountain
rose like a huge, rock-bound fortress from this plain-
like expanse. It was ledge upon ledge, precipice
upon precipice, up which and over which we made
our way slowly and with great labor, now pulling
ourselves up by our hands, then cautiously finding
niches for our feet and zigzagging right and left from
shelf to shelf. This northern side of the mountain
was thickly covered with moss and lichens, like the
north side of a tree. This made it soft to the foot,
and broke many a slip and fall. Everywhere a
stunted growth of yellow birch, mountain-ash, and
spruce and fir opposed our progress. The ascent at

such an angle with a roll of blankets on your back is not unlike climbing a tree: every limb resists your progress and pushes you back; so that when we at last reached the summit, after twelve or fifteen hundred feet of this sort of work, the fight was about all out of the best of us. It was then nearly two o'clock, so that we had been about seven hours in coming seven miles.

Here on the top of the mountain we overtook spring, which had been gone from the valley nearly a month. Red clover was opening in the valley below, and wild strawberries just ripening; on the summit the yellow birch was just hanging out its catkins, and the claytonia, or spring-beauty, was in bloom. The leaf-buds of the trees were just bursting, making a faint mist of green, which, as the eye swept downward, gradually deepened until it became a dense, massive cloud in the valleys. At the foot of the mountain the clintonia, or northern green lily, and the low shadbush were showing their berries, but long before the top was reached they were found in bloom. I had never before stood amid blooming claytonia, a flower of April, and looked down upon a field that held ripening strawberries. Every thousand feet elevation seemed to make about ten days' difference in the vegetation, so that the season was a month or more later on the top of the mountain than at its base. A very pretty flower which we began to meet with well up on the mountain-side was the painted trillium, the petals white, veined with pink.

The low, stunted growth of spruce and fir which clothes the top of Slide has been cut away over a small space on the highest point, laying open the view on nearly all sides. Here we sat down and enjoyed our triumph. We saw the world as the hawk or the balloonist sees it when he is three thousand feet in the air. How soft and flowing all the outlines of the hills and mountains beneath us looked! The forests dropped down and undulated away over them, covering them like a carpet. To the east we looked over the near-by Wittenberg range to the Hudson and beyond; to the south, Peak-o'-Moose, with its sharp crest, and Table Mountain, with its long level top, were the two conspicuous objects; in the west, Mt. Graham and Double Top, about three thousand eight hundred feet each, arrested the eye; while in our front to the north we looked over the top of Panther Mountain to the multitudinous peaks of the northern Catskills. All was mountain and forest on every hand. Civilization seemed to have done little more than to have scratched this rough, shaggy surface of the earth here and there. In any such view, the wild, the aboriginal, the geographical greatly predominate. The works of man dwindle, and the original features of the huge globe come out. Every single object or point is dwarfed; the valley of the Hudson is only a wrinkle in the earth's surface. You discover with a feeling of surprise that the great thing is the earth itself, which stretches away on every hand so far beyond your ken.

The Arabs believe that the mountains steady the earth and hold it together; but they have only to get on the top of a high one to see how insignificant mountains are, and how adequate the earth looks to get along without them. To the imaginative Oriental people, mountains seemed to mean much more than they do to us. They were sacred; they were the abodes of their divinities. They offered their sacrifices upon them. In the Bible, mountains are used as a symbol of that which is great and holy. Jerusalem is spoken of as a holy mountain. The Syrians were beaten by the Children of Israel because, said they, "their gods are gods of the hills; therefore were they stronger than we." It was on Mount Horeb that God appeared to Moses in the burning bush, and on Sinai that He delivered to him the law. Josephus says that the Hebrew shepherds never pasture their flocks on Sinai, believing it to be the abode of Jehovah. The solitude of mountain-tops is peculiarly impressive, and it is certainly easier to believe the Deity appeared in a burning bush there than in the valley below. When the clouds of heaven, too, come down and envelop the top of the mountain, — how such a circumstance must have impressed the old God-fearing Hebrews! Moses knew well how to surround the law with the pomp and circumstance that would inspire the deepest awe and reverence.

But when the clouds came down and enveloped us on Slide Mountain, the grandeur, the solemnity, were gone in a twinkling; the portentous-looking clouds proved to be nothing but base fog that wet us and extinguished the world for us. How tame, and prosy, and humdrum the scene instantly became! But when the fog lifted, and we looked from under it as from under a just-raised lid, and the eye plunged again like an escaped bird into those vast gulfs of space that opened at our feet, the feeling of grandeur and solemnity quickly came back.

The first want we felt on the top of Slide, after we had got some rest, was a want of water. Several of us cast about, right and left, but no sign of water was found. But water must be had, so we all started off deliberately to hunt it up. We had not gone many hundred yards before we chanced upon an ice-cave beneath some rocks, — vast masses of ice, with crystal pools of water near. This was good luck, indeed, and put a new and a brighter face on the situation.

Slide Mountain enjoys a distinction which no other mountain in the State, so far as is known, does — it has a thrush peculiar to itself. This thrush was discovered and described by Eugene P. Bicknell, of New York, in 1880, and has been named Bicknell's thrush. A better name would have been Slide Mountain thrush, as the bird so far has been found only on this mountain.[1] I did not see or hear it upon the Wittenberg, which is only a few miles distant, and only two hundred feet lower. In its appearance to the eye among the trees, one would not distinguish it from the gray-cheeked thrush of Baird, or the olive-backed thrush, but its song is totally different. The moment I heard it I said, "There is a new bird, a new thrush," for the quality of all thrush songs is the same. A moment more, and I knew it was Bicknell's thrush. The song is in a minor key, finer, more attenuated, and more under the breath than that of any other thrush. It seemed as if the bird was blowing in a delicate, slender, golden tube, so fine and yet so flute-like and resonant the song appeared. At times it was like a musical whisper of great sweetness and power. The birds were numerous about the summit, but we saw them nowhere else. No other thrush was seen, though a few times during our stay I caught a mere echo of the hermit's song far down the mountain-side. A bird I was not prepared to see or to hear was the black-poll warbler, a bird usually found much farther north, but here it was, amid the balsam firs, uttering its simple, lisping song.

[1] Bicknell's thrush turns out to be the more southern form of the gray-cheeked thrush, and is found on the higher mountains of New York and New England.

The rocks on the tops of these mountains are quite sure to attract one's attention, even if he have no eye for such things. They are masses of light reddish conglomerate, composed of round wave-worn quartz pebbles. Every pebble has been shaped and polished upon some ancient seacoast, probably the Devonian. The rock disintegrates where it is most exposed to the weather, and forms a loose sandy and pebbly soil. These rocks form the floor of the coal formation, but in the Catskill region only the floor remains; the superstructure has never existed, or has been swept away; hence one would look for a coal mine here over his head in the air, rather than under his feet.

This rock did not have to climb up here as we did; the mountain stooped and took it upon its back in the bottom of the old seas, and then got lifted up again. This happened so long ago that the memory of the oldest inhabitants of these parts yields no clew to the time.

A pleasant task we had in reflooring and reroofing the log-hut with balsam boughs against the night. Plenty of small balsams grew all about, and we soon had a huge pile of their branches in the old hut. What a transformation, this fresh green carpet and our fragrant bed, like the deep-furred robe of some huge animal, wrought in that dingy interior! Two or three things disturbed our sleep. A cup of strong beef-tea taken for supper disturbed mine; then the porcupines kept up such a grunting and chattering near our heads, just on the other side of the log, that sleep was difficult. In my wakeful mood I was a good deal annoyed by a little rabbit that kept whipping in at our dilapidated door and nibbling at our bread and hardtack. He persisted even after the gray of the morning appeared. Then about four o'clock it began gently to rain. I think I heard the first drop that fell. My companions were all in sound sleep. The rain increased, and gradually the sleepers awoke. It was like the tread of an advancing enemy which every ear had been expecting. The roof over us was of the poorest, and we had no confidence in it. It was made of the thin bark of spruce and balsam, and was full of hollows and depressions. Presently these hollows got full of water, when there was a simultaneous downpour of bigger and lesser rills upon the sleepers beneath. Said sleepers, as one man, sprang up, each taking his blanket with him; but by the time some of the party had got themselves stowed away under the adjacent rock, the rain ceased. It was little more than the dissolving of the nightcap of fog which so often hangs about these heights. With the first appearance of the dawn I had heard the new thrush in the scattered trees near the hut, — a strain as fine as if blown upon a fairy flute, a suppressed musical whisper from out the tops of the dark spruces. Probably never did there go up from the top of a great mountain a

smaller song to greet the day, albeit it was of the
purest harmony. It seemed to have in a more
marked degree the quality of interior reverberation
than any other thrush song I had ever heard. Would
the altitude or the situation account for its minor
key? Loudness would avail little in such a place.
Sounds are not far heard on a mountain-top; they
are lost in the abyss of vacant air. But amid these
low, dense, dark spruces, which make a sort of canopied privacy of every square rod of ground, what
could be more in keeping than this delicate musical
whisper? It was but the soft hum of the balsams,
interpreted and embodied in a bird's voice.

It was the plan of two of our companions to go
from Slide over into the head of the Rondout, and
thence out to the railroad at the little village of
Shokan, an unknown way to them, involving nearly
an all-day pull the first day through a pathless wilderness. We ascended to the topmost floor of the
tower, and from my knowledge of the topography
of the country I pointed out to them their course,
and where the valley of the Rondout must lie. The
vast stretch of woods, when it came into view from
under the foot of Slide, seemed from our point of
view very uniform. It swept away to the southeast,
rising gently toward the ridge that separates Lone
Mountain from Peak-o'-Moose, and presented a
comparatively easy problem. As a clew to the course,
the line where the dark belt or saddle-cloth of spruce,

which covered the top of the ridge they were to skirt, ended, and the deciduous woods began, a sharp, well-defined line was pointed out as the course to be followed. It led straight to the top of the broad level-backed ridge which connected two higher peaks, and immediately behind which lay the headwaters of the Rondout. Having studied the map thoroughly, and possessed themselves of the points, they rolled up their blankets about nine o'clock, and were off, my friend and I purposing to spend yet another day and night on Slide. As our friends plunged down into that fearful abyss, we shouted to them the old classic caution, "Be bold, be bold, *be not too* bold." It required courage to make such a leap into the unknown, as I knew those young men were making, and it required prudence. A faint heart or a bewildered head, and serious consequences might have resulted. The theory of a thing is so much easier than the practice! The theory is in the air, the practice is in the woods; the eye, the thought, travel easily where the foot halts and stumbles. However, our friends made the theory and the fact coincide ; they kept the dividing line between the spruce and the birches, and passed over the ridge into the valley safely; but they were torn and bruised and wet by the showers, and made the last few miles of their journey on will and pluck alone, their last pound of positive strength having been exhausted in making the descent through the chaos of rocks and logs into the head of the valley. In such emergencies one overdraws his account; he travels on the credit of the strength he expects to gain when he gets his dinner and some sleep. Unless one has made such a trip himself (and I have several times in my life), he can form but a faint idea what it is like, — what a trial it is to the body, and what a trial it is to the mind. You are fighting a battle with an enemy in ambush. How those miles and leagues which your feet must compass lie hidden there in that wilderness; how they seem to multiply themselves; how they are fortified with logs, and rocks, and fallen trees; how they take refuge in deep gullies, and skulk behind unexpected eminences! Your body not only feels the fatigue of the battle, your mind feels the strain of the undertaking; you may miss your mark; the mountains may outmanœuvre you. All that day, whenever I looked upon that treacherous wilderness, I thought with misgivings of those two friends groping their way there, and would have given much to know how it fared with them. Their concern was probably less than my own, because they were more ignorant of what was before them. Then there was just a slight shadow of a fear in my mind that I might have been in error about some points of the geography I had pointed out to them. But all was well, and the victory was won according to the campaign which I had planned. When we saluted our friends upon their own doorstep a week afterward, the wounds were nearly healed and the rents all mended.

When one is on a mountain-top, he spends most of the time in looking at the show he has been at such pains to see. About every hour we would ascend the rude lookout to take a fresh observation. With a glass I could see my native hills forty miles away to the northwest. I was now upon the back of the horse, yea, upon the highest point of his shoulders, which had so many times attracted my attention as a boy. We could look along his balsam-covered back to his rump, from which the eye glanced away down into the forests of the Never-sink, and on the other hand plump down into the gulf where his head was grazing or drinking. During the day there was a grand procession of thunder-clouds filing along over the northern Catskills, and letting down veils of rain and enveloping them. From such an elevation one has the same view of the clouds that he does from the prairie or the ocean. They do not seem to rest across and to be upborne by the hills, but they emerge out of the dim west, thin and vague, and grow and stand up as they get nearer and roll by him, on a level but invisible highway, huge chariots of wind and storm.

In the afternoon a thick cloud threatened us, but it proved to be the condensation of vapor that announces a cold wave. There was soon a marked fall in the temperature, and as night drew near it became pretty certain that we were going to have a cold time of it. The wind rose, the vapor above us thickened and came nearer, until it began to drive across the summit in slender wraiths, which curled over the brink and shut out the view. We became very diligent in getting in our night wood, and in gathering more boughs to calk up the openings in the hut. The wood we scraped together was a sorry lot, roots and stumps and branches of decayed spruce, such as we could collect without an axe, and some rags and tags of birch bark. The fire was built in one corner of the shanty, the smoke finding easy egress through large openings on the east side and in the roof over it. We doubled up the bed, making it thicker and more nest-like, and as darkness set in, stowed ourselves into it beneath our blankets. The searching wind found out every crevice about our heads and shoulders, and it was icy cold. Yet we fell asleep, and had slept about an hour when my companion sprang up in an unwonted state of excitement for so placid a man. His excitement was occasioned by the sudden discovery that what appeared to be a bar of ice was fast taking the place of his backbone. His teeth chattered, and he was convulsed with ague. I advised him to replenish the fire, and to wrap himself in his blanket and cut the liveliest capers he was capable of in so circumscribed a place. This he promptly did, and the thought of his wild and desperate dance there in the dim light, his tall form, his blanket flapping, his teeth chattering, the porcupines outside marking time with their squeals and grunts, still provokes a smile, though it was a serious enough matter at the time. After a while, the warmth came back to him, but he dared not trust himself again to the boughs; he fought the cold all night as one might fight a besieging foe. By carefully husbanding the fuel, the beleaguering enemy was kept at bay till morning came; but when morning did come, even the huge root he had used as a chair was consumed. Rolled in my blanket beneath a foot or more of balsam boughs, I had got some fairly good sleep, and was most of the time oblivious of the melancholy vigil of my friend. As we had but a few morsels of food left, and had been on rather short rations the day before, hunger was added to his other discomforts. At that time a letter was on the way to him from his wife, which contained this prophetic sentence: "I hope thee is not suffering with cold and hunger on some lone mountain-top."

Mr. Bicknell's thrush struck up again at the first signs of dawn, notwithstanding the cold. I could hear his penetrating and melodious whisper as I lay buried beneath the boughs. Presently I arose and invited my friend to turn in for a brief nap, while I gathered some wood and set the coffee brewing. With a brisk, roaring fire on, I left for the spring to fetch some water, and to make my toilet. The leaves of the mountain goldenrod, which everywhere covered the ground in the opening, were covered with frozen particles of vapor, and the scene, shut in by fog, was chill and dreary enough.

We were now not long in squaring an account with Slide, and making ready to leave. Round pellets of snow began to fall, and we came off the mountain on the 10th of June in a November storm and temperature. Our purpose was to return by the same valley we had come. A well-defined trail led off the summit to the north; to this we committed ourselves. In a few minutes we emerged at the head of the slide that had given the mountain its name. This was the path made by visitors to the scene; when it ended, the track of the avalanche began; no bigger than your hand, apparently, had it been at first, but it rapidly grew, until it became several rods in width. It dropped down from our feet straight as an arrow until it was lost in the fog, and looked perilously steep. The dark forms of the spruce were clinging to the edge of it, as if reaching out to their fellows to save them. We hesitated on the brink, but finally cautiously began the descent. The rock was quite naked and slippery, and only on the margin of the slide were there any boulders to stay the foot, or bushy growths to aid the hand. As we paused, after some minutes, to select our course, one of the finest surprises of the trip awaited us:

the fog in our front was swiftly whirled up by the breeze, like the drop-curtain at the theatre, only much more rapidly, and in a twinkling the vast gulf opened before us. It was so sudden as to be almost bewildering. The world opened like a book, and there were the pictures; the spaces were without a film, the forests and mountains looked surprisingly near; in the heart of the northern Catskills a wild valley was seen flooded with sunlight. Then the curtain ran down again, and nothing was left but the gray strip of rock to which we clung, plunging down into the obscurity. Down and down we made our way. Then the fog lifted again. It was Jack and his beanstalk renewed; new wonders, new views, awaited us every few moments, till at last the whole valley below us stood in the clear sunshine. We passed down a precipice, and there was a rill of water, the beginning of the creek that wound through the valley below; farther on, in a deep depression, lay the remains of an old snow-bank; Winter had made his last stand here, and April flowers were springing up almost amid his very bones. We did not find a palace, and a hungry giant, and a princess, at the end of our beanstalk, but we found a humble roof and the hospitable heart of Mrs. Larkins, which answered our purpose better. And we were in the mood, too, to have undertaken an eating-bout with any giant Jack ever discovered.

RIVERBY

Of all the retreats I have found amid the Catskills, there is no other that possesses quite so many charms for me as this valley, wherein stands Larkins's humble dwelling; it is so wild, so quiet, and has such superb mountain views. In coming up the valley, you have apparently reached the head of civilization a mile or more lower down; here the rude little houses end, and you turn to the left into the woods. Presently you emerge into a clearing again, and before you rises the rugged and indented crest of Panther Mountain, and near at hand, on a low plateau, rises the humble roof of Larkins, — you get a picture of the Panther and of the homestead at one glance. Above the house hangs a high, bold cliff covered with forest, with a broad fringe of blackened and blasted tree-trunks, where the cackling of the great pileated woodpecker may be heard; on the left a dense forest sweeps up to the sharp spruce-covered cone of the Wittenberg, nearly four thousand feet high, while at the head of the valley rises Slide over all. From a meadow just back of Larkins's barn, a view may be had of all these mountains, while the terraced side of Cross Mountain bounds the view immediately to the east. Running from the top of Panther toward Slide one sees a gigantic wall of rock, crowned with a dark line of fir. The forest abruptly ends, and in its stead rises the face of this colossal rocky escarpment, like some barrier built by the mountain gods. Eagles might nest here. It breaks the monotony of the world of woods very impressively.

I delight in sitting on a rock in one of these upper fields, and seeing the sun go down behind Panther. The rapid-flowing brook below me fills all the valley with a soft murmur. There is no breeze, but the great atmospheric tide flows slowly in toward the cooling forest; one can see it by the motes in the air illuminated by the setting sun: presently, as the air cools a little, the tide turns and flows slowly out. The long, winding valley up to the foot of Slide, five miles of primitive woods, how wild and cool it looks, its one voice the murmur of the creek! On the Wittenberg the sunshine lingers long; now it stands up like an island in a sea of shadows, then slowly sinks beneath the wave. The evening call of a robin or a veery at his vespers makes a marked impression on the silence and the solitude.

The following day my friend and I pitched our tent in the woods beside the stream where I had pitched it twice before, and passed several delightful days, with trout in abundance and wild strawberries at intervals. Mrs. Larkins's cream-pot, butter-jar, and bread-box were within easy reach. Near the camp was an unusually large spring, of icy coldness, which served as our refrigerator. Trout or milk immersed in this spring in a tin pail would keep sweet four or five days. One night some creature, probably a lynx or a raccoon, came and lifted the stone

from the pail that held the trout and took out a fine
string of them, and ate them up on the spot, leaving only the string and one head. In August bears
come down to an ancient and now brushy bark-
peeling near by for blackberries. But the creature
that most infests these backwoods is the porcupine.
He is as stupid and indifferent as the skunk; his
broad, blunt nose points a witless head. They are
great gnawers, and will gi aw youi house down if you
do not look out. Of a summer evening they will
walk coolly into your open door if not prevented.
The most annoying animal to the camper-out in this
region, and the one he needs to be most on the lookout for, is the cow. Backwoods cows and young
cattle seem always to be famished for salt, and they
will fairly lick the fisherman's clothes off his back,
and his tent and equipage out of existence, if you
give them a chance. On one occasion some wood-
ranging heifers and steers that had been hovering
around our camp for some days made a raid upon
it when we were absent. The tent was shut and
everything snugged up, but they ran their long
tongues under the tent, and, tasting something savory, hooked out John Stuart Mill's "Essays on Religion," which one of us had brought along, thinking to read in the woods. They mouthed the volume
around a good deal, but its logic was too tough for
them, and they contented themselves with devouring the paper in which it was wrapped. If the cat-

tle had not been surprised at just that point, it is probable the tent would have gone down before their eager curiosity and thirst for salt.

The raid which Larkins's dog made upon our camp was amusing rather than annoying. He was a very friendly and intelligent shepherd dog, probably a collie. Hardly had we sat down to our first lunch in camp before he called on us. But as he was disposed to be too friendly, and to claim too large a share of the lunch, we rather gave him the cold shoulder. He did not come again; but a few evenings afterward, as we sauntered over to the house on some trifling errand, the dog suddenly conceived a bright little project. He seemed to say to himself, on seeing us, "There come both of them now, just as I have been hoping they would; now, while they are away, I will run quickly over and know what they have got that a dog can eat." My companion saw the dog get up on our arrival, and go quickly in the direction of our camp, and he said something in the cur's manner suggested to him the object of his hurried departure. He called my attention to the fact, and we hastened back. On cautiously nearing camp, the dog was seen amid the pails in the shallow water of the creek investigating them. He had uncovered the butter, and was about to taste it, when we shouted, and he made quick steps for home, with a very "kill-sheep" look. When we again met him at the house next day, he could not look us in the face, but sneaked off, utterly crest-fallen. This was a clear case of reasoning on the part of the dog, and afterward a clear case of a sense of guilt from wrong-doing. The dog will probably be a man before any other animal.

IN MAMMOTH CAVE

*With President Theodore Roosevelt
in Yellowstone National Park in 1903*

EDITOR'S NOTE

Despite Burroughs's identification as the *genius loci* of the Hudson Valley and the Catskill Mountains, he traveled extensively outside New York State, particularly during the later years of his life. He visited England, France, Ireland, Canada, Jamaica, Bermuda, and Hawaii. In the eastern United States he traveled north to New England and south to Georgia and Florida. After traveling west to California and northwest to Alaska with the Harriman Expedition in 1899, he later visited Yellowstone with President Theodore Roosevelt in 1903 and traveled to the Grand Canyon and Yosemite with John Muir in 1909. On his travels he chafed at guided tours, and after three hours in Mammoth Cave during a trip to Kentucky in 1886, he wrote, "I am always more or less bored when I am toted about to see things which I am expected to admire." In the essay "In Mammoth Cave," published in the *Riverby* volume in 1894, Burroughs demonstrates how even as a tourist he relies on his own senses and his own particular experimental methods during his visit to the famous cave. In defining the significance of the cave's primordial darkness, where the objective universe is gone, Burroughs pays tribute to what he repeatedly calls "the light of day," a phrase that reappears as the title of a collection of his essays in 1900. "It is not for me to go about seeking wonders," Burroughs wrote after his visit to Mammoth Cave. "I do better along my old cow-paths."

F.B.

IN MAMMOTH CAVE

SOME idea of the impression which Mammoth Cave makes upon the senses, irrespective even of sight, may be had from the fact that blind people go there to see it, and are greatly struck with it. I was assured that this is a fact. The blind seem as much impressed by it as those who have their sight. When the guide pauses at the more interesting point, or lights the scene up with a great torch or with Bengal lights, and points out the more striking features, the blind exclaim, "How wonderful! how beautiful!" They can feel it, if they cannot see it. They get some idea of the spaciousness when words are uttered. The voice goes forth in these colossal chambers like a bird. When no word is spoken, the silence is of a kind never experienced on the surface of the earth, it is so profound and abysmal. This, and the absolute darkness, to a person with eyes makes him feel as if he were face to face with the primordial nothingness. The objective universe is gone; only the subjective remains; the sense of hearing is inverted, and reports only the murmurs from within. The blind miss much, but much remains to them.

The great cave is not merely a spectacle to the eye; it is a wonder to the ear, a strangeness to the smell and to the touch. The body feels the presence of unusual conditions through every pore.

For my part, my thoughts took a decidedly sepulchral turn; I thought of my dead and of all the dead of the earth, and said to myself, the darkness and the silence of their last resting-place is like this; to this we must all come at last. No vicissitudes of earth, no changes of seasons, no sound of storm or thunder penetrate here; winter and summer, day and night, peace or war, it is all one; a world beyond the reach of change, because beyond the reach of life. What peace, what repose, what desolation! The marks and relics of the Indian, which disappear so quickly from the light of day above, are here beyond the reach of natural change. The imprint of his moccasin in the dust might remain undisturbed for a thousand years. At one point the guide reaches his arm beneath the rocks that strew the floor and pulls out the burnt ends of canes, which were used, probably, when filled with oil or grease, by the natives to light their way into the cave doubtless centuries ago.

Here in the loose soil are ruts worn by cart-wheels in 1812, when, during the war with Great Britain, the earth was searched to make saltpetre. The guide kicks corn-cobs out of the dust where the oxen were fed at noon, and they look nearly as fresh as ever they did. In those frail corn-cobs and in those wheel-tracks as if the carts had but just gone along, one seemed to come very near to the youth of the century, almost to overtake it.

At a point in one of the great avenues, if you stop and listen, you hear a slow, solemn ticking like a great clock in a deserted hall; you hear the slight echo as it fathoms and sets off the silence. It is called the clock, and is caused by a single large drop of water falling every second into a little pool. A ghostly kind of clock there in the darkness, that is never wound up and that never runs down. It seemed like a mockery where time is not, and change does not come, — the clock of the dead. This sombre and mortuary cast of one's thoughts seems so natural in the great cave, that I could well understand the emotions of a lady who visited the cave with a party a few days before I was there. She went forward very reluctantly from the first; the silence and the darkness of the huge mausoleum evidently impressed her imagination, so that when she got to the spot where the guide points out the "Giant's Coffin," a huge, fallen rock, which in the dim light takes exactly the form of an enormous coffin, her fear quite overcame her, and she begged piteously to be taken back. Timid, highly imaginative people, especially women, are quite sure to have a sense of fear in this strange underground world. The guide told me of a lady in one of the parties he was conducting through, who wanted to linger behind a little all alone; he suffered her to do so, but presently heard a piercing scream. Rushing back, he found her lying prone upon the ground in a dead faint. She had accidentally put out her lamp, and was so appalled by the darkness that instantly closed around her that she swooned at once.

Sometimes it seemed to me as if I were threading the streets of some buried city of the fore-world. With your little lantern in your hand, you follow your guide through those endless and silent avenues, catching glimpses on either hand of what appears to be some strange antique architecture, the hoary and crumbling walls rising high up into the darkness. Now we turn a sharp corner, or turn down a street which crosses our course at right angles; now we come out into a great circle, or spacious court, which the guide lights up with a quick-paper torch, or a colored chemical light. There are streets above you and streets below you. As this was a city where day never entered, no provision for light needed to be made, and it is built one layer above another to the number of four or five, or on the plan of an enormous ant-hill, the lowest avenues being several hundred feet beneath the uppermost. The main avenue leading in from the entrance is called the Broadway, and if Broadway, New York, were arched over and reduced to utter darkness and silence, and its roadway blocked with mounds of earth and fragments of rock, it would, perhaps, only lack that gray, cosmic, elemental look, to make it resemble this. A mile or so from the entrance we pass a couple of rude stone houses, built forty or more years ago by some consumptives, who hoped to prolong their lives by a residence in this pure, antiseptic air. Five months they lived here, poor creatures, a half dozen of them, without ever going forth into the world of light. But the long entombment did not arrest the disease; the mountain did not draw the virus out, but seemed to draw the strength and vitality out, so that when the victims did go forth into the light and air, bleached as white as chalk, they succumbed at once, and nearly all died before they could reach the hotel, a few hundred yards away.

Probably the prettiest thing they have to show you in Mammoth Cave is the Star Chamber. This seems to have made an impression upon Emerson when he visited the cave, for he mentions it in one of his essays, "Illusions." The guide takes your lantern from you and leaves you seated upon a bench by the wayside, in the profound cosmic darkness. He retreats along a side alley that seems to go down to a lower level, and at a certain point shades his lamp with his hat, so that the light falls upon the ceiling over your head. You look up, and the first thought is that there is an opening just there that permits you to look forth upon the midnight skies. You see the darker horizon line where the sky ends and the mountains begin. The sky is blue-black and is thickly studded with stars, rather small stars, but apparently genuine. At one point a long, luminous streak simulates exactly the form and effect of a comet. As you gaze, the guide slowly moves his hat, and a black cloud gradually creeps over the sky, and all is blackness again. Then you hear footsteps retreating and dying away in the distance. Presently all is still, save the ringing in your own ears. Then after a few moments, during which you have sat in silence like that of the interstellar spaces, you hear over your left shoulder a distant flapping of wings, followed by the crowing of a cock. You turn your head in that direction and behold a faint dawn breaking on the horizon. It slowly increases till you hear footsteps approaching, and your dusky companion, playing the part of Apollo, with lamp in hand ushers in the light of day. It is rather theatrical, but a very pleasant diversion nevertheless.

Another surprise was when we paused at a certain point, and the guide asked me to shout or call in a loud voice. I did so without any unusual effect following. Then he spoke in a very deep bass, and instantly the rocks all about and beneath us became like the strings of an Æolian harp. They seemed transformed as if by enchantment. Then I tried, but did not strike the right key; the rocks were dumb; I tried again, but got no response; flat and dead the sounds came back as if in mockery; then

I struck a deeper bass, the chord was hit, and the solid walls seemed to become as thin and frail as a drum-head or as the frame of a violin. They fairly seemed to dance about us, and to recede away from us. Such wild, sweet music I had never before heard rocks discourse. Ah, the magic of the right key! "Why leap ye, ye high hills?" why, but that they had been spoken to in the right voice? Is not the whole secret of life to pitch our voices in the right key? Responses come from the very rocks when we do so. I thought of the lines of our poet of Democracy: —

"Surely, whoever speaks to me in the right voice, him or
    her I shall follow,
As the water follows the moon, silently, with fluid steps,
    anywhere around the globe."

Where we were standing was upon an arch over an avenue which crossed our course beneath us. The reverberations on Echo River, a point I did not reach, can hardly be more surprising, though they are described as wonderful.

There are four or five levels in the cave, and a series of avenues upon each. The lowest is some two hundred and fifty feet below the entrance. Here the stream which has done all this carving and tunneling has got to the end of its tether. It is here on a level with Green River in the valley below, and flows directly into it. I say the end of its tether, though if Green River cuts its valley deeper, the stream will, of course, follow suit. The bed of the river has probably, at successive periods, been on a level with each series of avenues of the cave. The stream is now doubtless but a mere fraction of its former self. Indeed, every feature of the cave attests the greater volume and activity of the forces which carved it, in the earlier geologic ages. The waters have worn the rock as if it were but ice. The domes and pits are carved and fluted in precisely the way dripping water flutes snow or ice. The rainfall must have been enormous in those early days, and it must have had a much stronger and sharper tooth of carbonic acid gas than now. It has carved out enormous pits with perpendicular sides, two or three hundred feet deep. Goring Dome I remember particularly. You put your head through an irregularly shaped window in the wall at the side of one of the avenues, and there is this huge shaft or well, starting from some higher level and going down two hundred feet below you. There must have been such wells in the old glaciers, worn by a rill of water slowly eating its way down. It was probably ten feet across, still moist and dripping. The guide threw down a lighted torch, and it fell and fell, till I had to crane my neck far out to see it finally reach the bottom. Some of these pits are simply appalling, and where the way is narrow, have been covered over to prevent accidents.

IN MAMMOTH CAVE

No part of Mammoth Cave was to me more impressive than its entrance, probably because here its gigantic proportions are first revealed to you, and can be clearly seen. That strange colossal underworld here looks out into the light of day, and comes in contrast with familiar scenes and objects. When you are fairly in the cave, you cannot see it; that is, with your aboveground eyes; you walk along by the dim light of your lamp as in a huge wood at night; when the guide lights up the more interesting portions with his torches and colored lights, the effect is weird and spectral; it seems like a dream; it is an unfamiliar world; you hardly know whether this is the emotion of grandeur which you experience, or of mere strangeness. If you could have the light of day in there, you would come to your senses, and could test the reality of your impressions. At the entrance you have the light of day, and you look fairly in the face of this underground monster, yea, into his open mouth, which has a span of fifty feet or more, and down into his contracting throat, where a man can barely stand upright, and where the light fades and darkness begins. As you come down the hill through the woods from the hotel, you see no sign of the cave till you emerge into a small opening where the grass grows and the sunshine falls, when you turn slightly to the right, and there at your feet yawns this terrible pit; and you feel indeed as if the mountain had opened its mouth and was lying in wait to swallow you down, as a whale might swallow a shrimp. I never grew tired of sitting or standing here by this entrance and gazing into it. It had for me something of the same fascination that the display of the huge elemental forces of nature have, as seen in thunder-storms, or in a roaring ocean surf. Two phœbe-birds had their nests in little niches of the rocks, and delicate ferns and wild flowers fringed the edges.

Another very interesting feature to me was the behavior of the cool air which welled up out of the mouth of the cave. It simulated exactly a fountain of water. It rose up to a certain level, or until it filled the depression immediately about the mouth of the cave, and then flowing over at the lowest point, ran down the hill toward Green River, along a little water-course, exactly as if it had been a liquid. I amused myself by wading down into it as into a fountain. The air above was muggy and hot, the thermometer standing at about eighty-six degrees, and this cooler air of the cave, which was at a temperature of about fifty-two degrees, was separated in the little pool or lakelet which is formed from the hotter air above it by a perfectly horizontal line. As I stepped down into it I could feel it close over my feet, then it was at my knees, then I was immersed to my hips, then to my waist, then I stood neck-deep in it, my body almost chilled, while my face and head were bathed by a sultry, oppressive air. Where the two bodies of air came in contact, a slight film of vapor was formed by condensation; I waded in till I could look under this as under a ceiling. It was as level and as well defined as a sheet of ice on a pond. A few moments' immersion into this aerial fountain made one turn to the warmer air again. At the depression in the rim of the basin one had but to put his hand down to feel the cold air flowing over like water. Fifty yards below you could still wade into it as into a creek, and at a hundred yards it was still quickly perceptible, but broader and higher; it had begun to lose some of its coldness, and to mingle with the general air; all the plants growing on the margin of the water-course were in motion, as well as the leaves on the low branches of the trees near by. Gradually this cool current was dissipated and lost in the warmth of the day.

WILD LIFE ABOUT MY CABIN

With his son Julian overlooking
Slabsides from "Julian's Rock"

EDITOR'S NOTE

In November 1895 Burroughs began building a cabin near a muck swamp on twenty acres of land he had acquired in the wooded hills west of his home. His son had discovered the secluded location while hunting partridge two years earlier. Burroughs completed the slab-sided cabin in early 1896 and contemplated calling it Echo Lodge, Foot Cliffs, Rock Haven, Whippoorwill's Nest, Coon Hollow, among other names, before he took a friend's suggestion and called it simply Slabsides, in reference to the slabs of hemlock and chestnut covering the outer walls. In "Wild Life about My Cabin," Burroughs explains the meaning of the name "Slabsides" as well as the significance of the cabin and its location. For the next twenty-five years the rustic retreat served as a place for Burroughs to write and as a guest house. In the area around the cabin, marked by Black Creek, Black Pond, and Julian's Rock (a high rock with a "grand view of the West" named in honor of his son), Burroughs found a country more wild and yet more intimate than that along the Hudson River. While drifting

EDITOR'S NOTE

across Black Pond in a canoe he enters a Thoreauvian series of reflections on the nature of ponds and hunting. The essay, which Burroughs wrote while visiting his son at Harvard in 1898, was first published as "Glimpses of Wild Life About My Cabin" in *Century* in August 1899 and was reprinted under its present title in *Far and Near* in 1904.

F.B.

WILD LIFE ABOUT MY CABIN

FRIENDS have often asked me why I turned my back upon the Hudson and retreated into the wilderness. Well, I do not call it a retreat; I call it a withdrawal, a retirement, the taking up of a new position to renew the attack, it may be, more vigorously than ever. It is not always easy to give reasons. There are reasons within reasons, and often no reasons at all that we are aware of.

To a countryman like myself, not born to a great river or an extensive water-view, these things, I think, grow wearisome after a time. He becomes surfeited with a beauty that is alien to him. He longs for something more homely, private, and secluded. Scenery may be too fine or too grand and imposing for one's daily and hourly view. It tires after a while. It demands a mood that comes to you only at intervals. Hence it is never wise to build your house on the most ambitious spot in the landscape. Rather seek out a more humble and secluded nook or corner, which you can fill and warm with your domestic and home instincts and affections. In some things the half is often more satisfying than the whole. A glimpse of the Hudson River between hills or through openings in the trees wears better with me than a long expanse of it constantly spread out before me. One day I had an errand to a farmhouse nestled in a little valley or basin at the foot of a mountain. The earth put out protecting arms all about it, — a low hill with an orchard on one side, a sloping pasture on another, and the mountain, with the skirts of its mantling forests, close at hand in the rear. How my heart warmed toward it! I had been so long perched high upon the banks of a great river, in sight of all the world, exposed to every wind that blows, with a horizon-line that sweeps over half a county, that, quite unconsciously to myself, I was pining for a nook to sit down in. I was hungry for the private and the circumscribed; I knew it when I saw this sheltered farmstead. I had long been restless and dissatisfied, — a vague kind of homesickness; now I knew the remedy. Hence when, not long afterward, I was offered a tract of wild land, barely a mile from home, that contained a secluded nook and a few acres of level, fertile land shut off from the vain and noisy world of railroads, steamboats, and yachts by a wooded, precipitous mountain, I quickly closed the bargain, and built me a rustic house there, which I call "Slabsides," because its outer walls are covered with slabs. I might have given it a prettier name, but not one more fit, or more in keeping with the mood that brought me thither.

WILD LIFE ABOUT MY CABIN

A slab is the first cut from the log, and the bark goes
with it. It is like the first cut from the loaf, which
we call the crust, and which the children reject, but
which we older ones often prefer. I wanted to take a
fresh cut of life, — something that had the bark on,
or, if you please, that was like a well-browned and
hardened crust. After three years I am satisfied with
the experiment. Life has a different flavor here. It
is reduced to simpler terms; its complex equations
all disappear. The exact value of x may still elude
me, but I can press it hard; I have shorn it of
many of its disguises and entanglements.

When I went into the woods the robins went with
me, or rather they followed close. As soon as a space
of ground was cleared and the garden planted, they
were on hand to pick up the worms and insects, and
to superintend the planting of the cherry-trees: three
pairs the first summer, and more than double that
number the second. In the third, their early morning chorus was almost as marked a feature as it is
about the old farm homesteads. The robin is no hermit: he likes company; he likes the busy scenes of
the farm and the village; he likes to carol to listening ears, and to build his nest as near your dwelling
as he can. Only at rare intervals do I find a real
sylvan robin, one that nests in the woods, usually by
still waters, remote from human habitation. In such
places his morning and evening carol is a welcome
surprise to the fisherman or camper-out. It is like a

dooryard flower found blooming in the wilderness. With the robins came the song sparrows and social sparrows, or chippies, also. The latter nested in the bushes near my cabin, and the song sparrows in the bank above the ditch that drains my land. I notice that Chippy finds just as many horsehairs to weave into her nest here in my horseless domain as she does when she builds in the open country. Her partiality for the long hairs from the manes and tails of horses and cattle is so great that she is often known as the hair-bird. What would she do in a country where there were neither cows nor horses? Yet these hairs are not good nesting-material. They are slippery, refractory things, and occasionally cause a tragedy in the nest by getting looped around the legs or the neck of the young or of the parent bird. They probably give a smooth finish to the interior, dear to the heart of Chippy.

The first year of my cabin life a pair of robins attempted to build a nest upon the round timber that forms the plate under my porch roof. But it was a poor place to build in. It took nearly a week's time and caused the birds a great waste of labor to find this out. The coarse material they brought for the foundation would not bed well upon the rounded surface of the timber, and every vagrant breeze that came along swept it off. My porch was kept littered with twigs and weed-stalks for days, till finally the birds abandoned the undertaking. The next season a wiser or more experienced pair made the attempt again, and succeeded. They placed the nest against the rafter where it joins the plate; they used mud from the start to level up with and to hold the first twigs and straws, and had soon completed a firm, shapely structure. When the young were about ready to fly, it was interesting to note that there was apparently an older and a younger, as in most families. One bird was more advanced than any of the others. Had the parent birds intentionally stimulated it with extra quantities of food, so as to be able to launch their offspring into the world one at a time? At any rate, one of the birds was ready to leave the nest a day and a half before any of the others. I happened to be looking at it when the first impulse to get outside the nest seemed to seize it. Its parents were encouraging it with calls and assurances from some rocks a few yards away. It answered their calls in vigorous, strident tones. Then it climbed over the edge of the nest upon the plate, took a few steps forward, then a few more, till it was a yard from the nest and near the end of the timber, and could look off into free space. Its parents apparently shouted, "Come on!" But its courage was not quite equal to the leap; it looked around, and seeing how far it was from home, scampered back to the nest, and climbed into it like a frightened child. It had made its first journey into the world, but the home tie had brought it quickly back.

A few hours afterward it journeyed to the end of the plate again, and then turned and rushed back. The third time its heart was braver, its wings stronger, and leaping into the air with a shout, it flew easily to some rocks a dozen or more yards away. Each of the young in succession, at intervals of nearly a day, left the nest in this manner. There would be the first journey of a few feet along the plate, the first sudden panic at being so far from home, the rush back, a second and perhaps a third attempt, and then the irrevocable leap into the air, and a clamorous flight to a near-by bush or rock. Young birds never go back when they have once taken flight. The first free flap of the wing severs forever the ties that bind them to home.

The chickadees we have always with us. They are like the evergreens among the trees and plants. Winter has no terrors for them. They are properly wood-birds, but the groves and orchards know them also. Did they come near my cabin for better protection, or did they chance to find a little cavity in a tree there that suited them? Branch-builders and ground-builders are easily accommodated, but the chickadee must find a cavity, and a small one at that. The woodpeckers make a cavity when a suitable trunk or branch is found, but the chickadee, with its small, sharp beak, rarely does so; it usually smooths and deepens one already formed. This a pair did a few yards from my cabin. The opening was into the heart of a little sassafras, about four feet from the ground. Day after day the birds took turns in deepening and enlarging the cavity: a soft, gentle hammering for a few moments in the heart of the little tree, and then the appearance of the worker at the opening, with the chips in his, or her, beak. They changed off every little while, one working while the other gathered food. Absolute equality of the sexes, both in plumage and in duties, seems to prevail among these birds, as among a few other species. During the preparations for housekeeping the birds were hourly seen and heard, but as soon as the first egg was laid, all this was changed. They suddenly became very shy and quiet. Had it not been for the new egg that was added each day, one would have concluded that they had abandoned the place. There was a precious secret now that must be well kept. After incubation began, it was only by watching that I could get a glimpse of one of the birds as it came quickly to feed or to relieve the other.

One day a lot of Vassar girls came to visit me, and I led them out to the little sassafras to see the chickadees' nest. The sitting bird kept her place as head after head, with its nodding plumes and millinery, appeared above the opening to her chamber, and a pair of inquisitive eyes peered down upon her. But I saw that she was getting ready to play her little trick to frighten them away. Presently I heard a faint explosion at the bottom of the cavity, when the peeping girl jerked her head quickly back, with the exclamation, "Why, it spit at me!" The trick of the bird on such occasions is apparently to draw in its breath till its form perceptibly swells, and then give forth a quick, explosive sound like an escaping jet of steam. One involuntarily closes his eyes and jerks back his head. The girls, to their great amusement, provoked the bird into this pretty outburst of her impatience two or three times. But as the ruse failed of its effect, the bird did not keep it up, but let the laughing faces gaze till they were satisfied.

There is only one other bird known to me that resorts to the same trick to scare away intruders, and that is the great crested flycatcher. As your head appears before the entrance to the cavity in which the mother bird is sitting, a sudden burst of escaping steam seems directed at your face, and your backward movement leaves the way open for the bird to escape, which she quickly does.

The chickadee is a prolific bird, laying from six to eight eggs, and it seems to have few natural enemies. I think it is seldom molested by squirrels or black snakes or weasels or crows or owls. The entrance to the nest is usually so small that none of these creatures can come at them. Yet the number of chickadees in any given territory seems small. What keeps them in check? Probably the rigors of winter and a limited food-supply. The ant-eaters, fruit-eaters, and seed-eaters mostly migrate. Our all-the-year-round birds, like the chickadees, woodpeckers, jays, and nuthatches, live mostly on nuts and the eggs and larvæ of tree-insects, and hence their larder is a restricted one; hence, also, these birds rear only one brood in a season. A hairy woodpecker passed the winter in the woods near me by subsisting on a certain small white grub which he found in the bark of some dead hemlock-trees. He "worked" these trees, — four of them, — as the slang is, "for all they were worth." The grub was under the outer shell of bark, and the bird literally skinned the trees in getting at his favorite morsel. He worked from the top downward, hammering or prying off this shell, and leaving the trunk of the tree with a red, denuded look. Bushels of the fragments of the bark covered the ground at the foot of the tree in spring, and the trunk looked as if it had been flayed, — as it had.

The big chimney of my cabin of course attracted the chimney swifts, and as it was not used in summer, two pairs built their nests in it, and we had the muffled thunder of their wings at all hours of the day and night. One night, when one of the broods was nearly fledged, the nest that held them fell down into the fireplace. Such a din of screeching and chattering as they instantly set up! Neither my dog nor I could sleep. They yelled in chorus, stopping at the end of every half-minute as if upon signal. Now they were all screeching at the top of their voices, then a sudden, dead silence ensued. Then the din began again, to terminate at the instant as before. If they had been long practicing together, they could not have succeeded better. I never before heard the cry of birds so accurately timed. After a while I got up and put them back up the chimney, and stopped up the throat of the flue with newspapers. The next day one of the parent birds, in bringing food to them, came down the chimney with such force that it passed through the papers and brought up in the fireplace. On capturing it I saw that its throat was distended with food as a chipmunk's cheek with corn, or a boy's pocket with chestnuts. I opened its mandibles, when it ejected a wad of insects as large as a bean. Most of them were much macerated, but there were two house-flies yet alive and but little the worse for their close confinement. They stretched themselves, and walked about upon my hand, enjoying a breath of fresh air once more. It was nearly two hours before the swift again ventured into the chimney with food.

These birds do not perch, nor alight upon buildings or the ground. They are apparently upon the wing all day. They outride the storms. I have in my mind a cheering picture of three of them I saw facing a heavy thunder-shower one afternoon. The wind was blowing a gale, the clouds were rolling in black, portentous billows out of the west, the peals of thunder were shaking the heavens, and the big drops were just beginning to come down, when, on looking up, I saw three swifts high in air, working their way slowly, straight into the teeth of the storm. They were not hurried or disturbed; they held themselves firmly and steadily; indeed, they were fairly at anchor in the air till the rage of the elements should have subsided. I do not know that any other of our land birds outride the storms in this way.

The phœbe-birds also soon found me out in my retreat, and a pair of them deliberated a long while about building on a little shelf in one of my gables. But, much to my regret, they finally decided in favor of a niche in the face of a ledge of rocks not far from my spring. The place was well screened by bushes and well guarded against the approach of snakes or four-footed prowlers, and the birds prospered well and reared two broods. They have now occupied the same nest three years in succession. This is unusual: Phœbe prefers a new nest each season, but in this case there is no room for another, and, the site being a choice one, she slightly repairs and refurnishes her nest each spring, leaving the new houses for her more ambitious neighbors.

Of wood-warblers my territory affords many specimens. One spring a solitary Nashville warbler lingered near my cabin for a week. I heard his bright, ringing song at all hours of the day. The next spring there were two or more, and they nested in my pea-bushes. The black and white creeping warblers are perhaps the most abundant. A pair of them built a nest in a steep moss and lichen covered hillside, beside a high gray rock. Our path to Julian's Rock led just above it. It was an ideal spot and an ideal nest, but it came to grief. Some small creature sucked the eggs. On removing the nest I found an earth-stained egg beneath it. Evidently the egg had ripened before its receptacle was ready, and the mother, for good luck, had placed it in the foundation.

One day, as I sat at my table writing, I had a call from the worm-eating warbler. It came into the open door, flitted about inquisitively, and then, startled by the apparition at the table, dashed against the window-pane and fell down stunned. I picked it up, and it lay with closed eyes panting in my hand. I carried it into the open air. In a moment or two it opened its eyes, looked about, and then closed them and fell to panting again. Soon it looked up at me once more and about the room, and seemed to say: "Where am I? What has happened to me?" Presently the panting ceased, the bird's breathing became more normal, it gradually got its bearings, and, at a motion of my hand, darted away. This is an abundant warbler in my vicinity, and nested this year near by. I have discovered that it has an air-song — the song of ecstasy — like that of the oven-bird. I had long suspected it, as I frequently heard a fine burst of melody that was new to me. One June day I was fortunate enough to see the bird delivering its song in the air above the low trees. As with the oven-bird, its favorite hour is the early twilight, though I hear the song occasionally at other hours. The bird darts upward fifty feet or more, about half the height that the oven-bird attains, and gives forth a series of rapid, ringing musical notes, which quickly glide into the long, sparrow-like trill that forms its ordinary workaday song. While this part is being uttered, the singer is on its downward flight into the woods. The flight-song of the oven-bird is louder and more striking, and is not so shy and furtive a performance. The latter I hear many times every June twilight, and I frequently see the singer reach his climax a hundred feet or more in the air, and then mark his arrow-like flight downward. I have heard this song also in the middle of the night near my cabin. At such times it stands out on the stillness like a bursting rocket on the background of the night.

One or two mornings in April, at a very early hour, I am quite sure to hear the hermit thrush singing in the bushes near my window. How quickly I am transported to the Delectable Mountains and to the mossy solitudes of the northern woods! The winter wren also pauses briefly in his northern journey, and surprises and delights my ear with his sudden lyrical burst of melody. Such a dapper, fidgety, gesticulating, bobbing-up-and-down-and-out-and-in little bird, and yet full of such sweet, wild melody! To get him at his best, one needs to hear him in a dim, northern hemlock wood, where his voice reverberates as in a great hall; just as one should hear the veery in a beech and birch wood, beside a purling trout brook, when the evening shades are falling. It then becomes to you the voice of some particular spirit of the place and the hour. The veery does not inhabit the woods immediately about my cabin, but in the summer twilight he frequently comes up from the valley below and sings along the borders of my territory. How welcome his simple flute-like strain! The wood thrush is the leading chorister in the woods about me. He does not voice the wildness, but seems to give a touch of something half rural, half urban, — such is the power of association in bird-songs. In the evening twilight I often sit on the highest point of the rocky rim of the great granite bowl that holds my three acres of prairie soil, and see the shadows deepen, and listen to the bird voices that rise up from the forest below me. The songs of many wood thrushes make a sort of golden warp in the texture of sounds that is being woven about me. Now the flight-song of the oven-bird holds the ear, then the fainter one of the worm-eating warbler lures it. The carol of the robin, the vesper hymn of the tanager, the flute of the veery, are all on the air. Finally, as the shadows deepen and the stars begin to come out, the whip-poor-will suddenly strikes up. What a rude intrusion upon the serenity and harmony of the hour! A cry without music, insistent, reiterated, loud, penetrating, and yet the ear welcomes it also; the night and the solitude are so vast that they can stand it; and when, an hour later, as the night enters into full possession, the bird comes and serenades me under my window or upon my doorstep, my heart warms toward it. Its cry is a love-call, and there is something of the ardor and persistence of love in it, and when the female responds, and comes and hovers near, there is an interchange of subdued, caressing tones between the two birds that it is a delight to hear. During my first summer here one bird used to strike up every night from a high ledge of rocks in front of my door. At just such a moment in the twilight he would begin, the first to break the stillness. Then the others would follow, till the solitude was vocal with their calls. They are rarely heard later than ten o'clock. Then at day-break they take up the tale again, whipping poor Will till one pities him. One April morning between three and four o'clock, hearing one strike up near my window, I began counting its calls. My neighbor had told me he had heard one call over two hundred times without a break, which seemed to me a big story. But I have a much bigger one to tell. This bird actually laid upon the back of poor Will one thousand and eighty-eight blows, with only a barely perceptible pause here and there, as if to catch its breath. Then it stopped about half a minute and began again, uttering this time three hundred and ninety calls, when it paused, flew a little farther away, took up the tale once more, and continued till I fell asleep.

By day the whip-poor-will apparently sits motionless upon the ground. A few times in my walks through the woods I have started one up from almost under my feet. On such occasions the bird's movements suggest those of a bat; its wings make no noise, and it wavers about in an uncertain manner, and quickly drops to the ground again. One June day we flushed an old one with her two young, but there was no indecision or hesitation in the manner of the mother bird this time. The young were more than half fledged, and they scampered away a few yards and suddenly squatted upon the ground, where their protective coloring rendered them almost invisible. Then the anxious parent put forth all her arts to absorb our attention and lure us away from her offspring. She flitted before us from side to side, with spread wings and tail, now falling upon the ground, where she would remain a moment as if quite disabled, then perching upon an old stump or low branch with drooping, quivering wings, and imploring us by every gesture to take her and spare her young. My companion had his camera with him, but the bird would not remain long enough in one position for him to get her picture. The whip-poor-will builds no nest, but lays her two blunt, speckled eggs upon the dry leaves, where the plumage of the sitting bird blends perfectly with her surroundings. The eye, only a few feet away, has to search long and carefully to make her out. Every gray and brown and black tint of dry leaf and lichen, and bit of bark or broken twig, is copied in her plumage. In a day or two, after the young are hatched, the mother begins to move about with them through the woods.

When I want the wild of a little different flavor and quality from that immediately about my cabin, I go a mile through the woods to Black Creek, here called the Shattega, and put my canoe into a long, smooth, silent stretch of water that winds through a heavily timbered marsh till it leads into Black Pond, an oval sheet of water half a mile or more across. Here I get the moist, spongy, tranquil, luxurious side of Nature. Here she stands or sits knee-deep in water, and wreathes herself with pond-lilies in summer, and bedecks herself with scarlet maples in autumn. She is an Indian maiden, dark, subtle, dreaming, with glances now and then that thrill the wild blood in one's veins. The Shattega here is a stream without banks and with a just perceptible current. It is a waterway through a timbered marsh. The level floor of the woods ends in an irregular line

where the level surface of the water begins. As one
glides along in his boat, he sees various rank aqua
tic growths slowly waving in the shadowy depths
beneath him. The larger trees on each side unite
their branches above his head, so that at times he
seems to be entering an arboreal cave out of which
glides the stream. In the more open places the
woods mirror themselves in the glassy surface till
one seems floating between two worlds, clouds and
sky and trees below him matching those around and
above him. A bird flits from shore to shore, and one
sees it duplicated against the sky in the under-world.
What vistas open! What banks of drooping foliage,
what grain and arch of gnarled branches, lure the
eye as one drifts or silently paddles along! The
stream has absorbed the shadows so long that it is
itself like a liquid shadow. Its bed is lined with various dark vegetable growths, as with the skin of
some huge, shaggy animal, the fur of which slowly
stirs in the languid current. I go here in early spring,
after the ice has broken up, to get a glimpse of the
first wild ducks and to play the sportsman without a
gun. I am sure I would not exchange the quiet surprise and pleasure I feel, as, on rounding some point
or curve in the stream, two or more ducks spring
suddenly out from some little cove or indentation in
the shore, and with an alarum *quack*, *quack*, launch
into the air and quickly gain the free spaces above
the treetops, for the satisfaction of the gunner who

sees their dead bodies fall before his murderous fire.
He has only a dead duck, which, the chances are, he
will not find very toothsome at this season, while I
have a live duck with whistling wings cleaving the
air northward, where, in some lake or river of Maine
or Canada, in late summer, I may meet him again
with his brood. It is so easy, too, to bag the game
with your eye, while your gun may leave you only a
feather or two floating upon the water. The duck
has wit, and its wit is as quick as, or quicker than, the
sportsman's gun. One day in spring I saw a gunner
cut down a duck when it had gained an altitude of
thirty or forty feet above the stream. At the report
it stopped suddenly, turned a somersault, and fell
with a splash into the water. It fell like a brick, and
disappeared like one; only a feather and a few bubbles marked the spot where it struck. Had it sunk?
No; it had dived. It was probably winged, and in
the moment it occupied in falling to the water it had
decided what to do. It would go beneath the hunter,
since it could not escape above him; it could fly in
the water with only one wing, with its feet to aid it.
The gunner instantly set up a diligent search in all
directions, up and down along the shores, peering
long and intently into the depths, thrusting his oar
into the weeds and driftwood at the edge of the water, but no duck or sign of duck could he find. It
was as if the wounded bird had taken to the mimic
heaven that looked so sunny and real down there,

and gone on to Canada by that route. What astonished me was that the duck should have kept its presence of mind under such trying circumstances, and not have lost a fraction of a second of time in deciding on a course of action. The duck, I am convinced, has more sagacity than any other of our commoner fowl.

The day I see the first ducks I am pretty sure to come upon the first flock of blackbirds, — rusty grackles, — resting awhile on their northward journey amid the reeds, alders, and spice-bush beside the stream. They allow me to approach till I can see their yellow eyes and the brilliant iris on the necks and heads of the males. Many of them are vocal, and their united voices make a volume of sound that is analogous to a bundle of slivers. Sputtering, splintering, rasping, rending, their notes chafe and excite the ear. They suggest thorns and briers of sound, and yet are most welcome. What voice that rises from our woods or beside our waters in April is not tempered or attuned to the ear? Just as I like to chew the crinkleroot and the twigs of the spice-bush at this time, or at any time, for that matter, so I like to treat my ear to these more aspirated and astringent bird voices. Is it Thoreau who says they are like pepper and salt to this sense? In all the blackbirds we hear the voice of April not yet quite articulate; there is a suggestion of catarrh and influenza still in the air-passages. I should, perhaps, except the red-shouldered starling, whose clear and liquid *gur-ga-lee* or *o-ka-lee*, above the full water-courses, makes a different impression. The cowbird also has a clear note, but it seems to be wrenched or pumped up with much effort.

In May I go to Black Creek to hear the warblers and the water-thrushes. It is the only locality where I have ever heard the two water-thrushes, or accentors, singing at the same time, — the New York and the large-billed. The latter is much more abundant and much the finer songster. How he does make these watery solitudes ring with his sudden, brilliant burst of song! But the more northern species pleases the ear also with his quieter and less hurried strain. I drift in my boat and let the ear attend to the one, then to the other, while the eye takes note of their quick, nervous movements and darting flight. The smaller species probably does not nest along this stream, but the large-billed breeds here abundantly. The last nest I found was in the roots of an upturned tree, with the water immediately beneath it. I had asked a neighboring farm-boy if he knew of any birds' nests.

"Yes," he said; and he named over the nests of robins, highholes, sparrows, and others, and then that of a "tip-up."

At this last I pricked up my ears, so to speak. I had not seen a tip-up's nest in many a day. "Where?" I inquired.

"In the roots of a tree in the woods," said Charley.

"Not the nest of the 'tip-up,' or sandpiper," said I. "It builds on the ground in the open country near streams."

"Anyhow, it tipped," replied the boy.

He directed me to the spot, and I found, as I expected to find, the nest of the water-thrush. When the Vassar girls came again, I conducted them to the spot, and they took turns in walking a small tree trunk above the water, and gazing upon a nest brimming with the downy backs of young birds.

When I am listening to the water-thrushes, I am also noting with both eye and ear the warblers and vireos. There comes a week in May when the speckled Canada warblers are in the ascendant. They feed in the low bushes near the water's edge, and are very brisk and animated in voice and movement. The eye easily notes their slate-blue backs and yellow breasts with their broad band of black spots, and the ear quickly discriminates their not less marked and emphatic song.

In late summer I go to the Shattega, and to the lake out of which it flows, for white pond-lilies, and to feast my eye on the masses of purple loosestrife and the more brilliant but more hidden and retired cardinal-flower that bloom upon its banks. One cannot praise the pond-lily; his best words mar it, like the insects that eat its petals: but he can contemplate it as it opens in the morning sun and distills such perfume, such purity, such snow of petal and such gold of anther, from the dark water and still darker ooze. How feminine it seems beside its coarser and more robust congeners; how shy, how pliant, how fine in texture and star-like in form!

The loosestrife is a foreign plant, but it has made itself thoroughly at home here, and its masses of royal purple make the woods look civil and festive. The cardinal burns with a more intense fire, and fairly lights up the little dark nooks where it glasses itself in the still water. One must pause and look at it. Its intensity, its pure scarlet, the dark background upon which it is projected, its image in the still darker water, and its general air of retirement and seclusion, all arrest and delight the eye. It is a heart-throb of color on the bosom of the dark solitude.

The rarest and wildest animal that my neighborhood boasts of is the otter. Every winter we see the tracks of one or more of them upon the snow along Black Creek. But the eye that has seen the animal itself in recent years I cannot find. It probably makes its excursions along the creek by night. Follow its track — as large as that of a fair-sized dog — over the ice, and you will find that it ends at every open pool and rapid, and begins again upon the ice beyond. Sometimes it makes little excursions up the bank, its body often dragging in the snow like a log. My son followed the track one day far up the mountain-side, where the absence of the snow caused him to lose it. I like to think of so wild and shy a creature holding its own within sound of the locomotive's whistle.

The fox passes my door in winter, and probably in summer too, as do also the 'possum and the coon. The latter tears down my sweet corn in the garden, and the rabbit eats off my raspberry-bushes and nibbles my first strawberries, while the woodchucks eat my celery and beans and peas. Chipmunks carry off the corn I put out for the chickens, and weasels eat the chickens themselves.

Many times during the season I have in my solitude a visit from a bald eagle. There is a dead tree near the summit, where he often perches, and which we call the " old eagle-tree." It is a pine, killed years ago by a thunderbolt, — the bolt of Jove, — and now the bird of Jove hovers about it or sits upon it. I have little doubt that what attracted me to this spot attracts him, — the seclusion, the savageness, the elemental grandeur. Sometimes, as I look out of my window early in the morning, I see the eagle upon his perch, preening his plumage, or waiting for the rising sun to gild the mountain-tops. When the smoke begins to rise from my chimney, or he sees me going to the spring for water, he concludes it is time for him to be off. But he need not fear the crack of the rifle here ; nothing more deadly than field-glasses shall be pointed at him while I am about. Often in the course of the day I see him circling above my domain, or winging his way toward the mountains. His home is apparently in the Shawangunk Range, twenty or more miles distant, and I fancy he stops or lingers above me on his way to the river. The days on which I see him are not quite the same as the other days. I think my thoughts soar a little higher all the rest of the morning: I have had a visit from a messenger of Jove. The lift or range of those great wings has passed into my thought. I once heard a collector get up in a scientific body and tell how many eggs of the bald eagle he had clutched that season, how many from this nest, how many from that, and how one of the eagles had deported itself after he had killed its mate. I felt ashamed for him. He had only proved himself a superior human weasel. The man with the rifle and the man with the collector's craze are fast reducing the number of eagles in the country. Twenty years ago I used to see a dozen or more along the river in the spring when the ice was breaking up, where I now see only one or two, or none at all. In the present case, what would it profit me could I find and plunder my eagle's nest, or strip his skin from his dead carcass? Should I know him better? I do not want to know him that way. I want rather to feel the inspiration of his presence and noble bearing. I want my interest and sympathy to go with him in his continental voyaging

up and down, and in his long, elevated flights to and
from his eyrie upon the remote, solitary cliffs. He
draws great lines across the sky ; he sees the forests
like a carpet beneath him, he sees the hills and valleys as folds and wrinkles in a many-colored tapes-
try ; he sees the river as a silver belt connecting remote horizons. We climb mountain-peaks to get a
glimpse of the spectacle that is hourly spread out
beneath him. Dignity, elevation, repose, are his. I
would have my thoughts take as wide a sweep. I
would be as far removed from the petty cares and
turmoils of this noisy and blustering world.

AUGUST DAYS

At Black Creek in the woods near Slabsides

EDITOR'S NOTE

In 1894 Burroughs announced to his readers that
the volume *Riverby* would probably be his last
collection of outdoor essays. In his journal he
confided that his interest, his curiosity, were blunted.
His growing inclination was toward philosophical
and literary themes. Ten years later in the preface
to *Far and Near* (1904), he stated that contrary to
his previous announcement, his love of nature had
continued, his habit of observation had been kept
up, and the result was another collection of essays
"dealing with the old inexhaustible, open-air themes."
In fact, Burroughs suggested, there would be future
volumes of such essays. In *Far and Near,* he shows
that despite his interest in faraway places like
Jamaica and Alaska, his primary attraction is still
to the natural world near to home. In "August
Days," he presents a dramatic personification of
the Hudson Valley in August as tanned, hirsute,
and freckled. The aesthetic values of the essay
form an interesting comparison with those of "April,"
published twenty-seven years earlier. Like many

EDITOR'S NOTE

American naturalists since the time of John and William Bartram, Burroughs shows an inclination toward phenological observations. "August Days" demonstrates his abiding concern with natural relationships as he dramatizes the connections of organisms to time as well as place.

F.B.

AUGUST DAYS

ONE of our well-known poets, in personifying August, represents her as coming with daisies in her hair. But an August daisy is a sorry affair; it is little more than an empty, or partly empty, seed-vessel. In the Northern States the daisy is in her girlhood and maidenhood in June; she becomes very matronly early in July, — fat, faded, prosaic, — and by or before August she is practically defunct. I recall no flower whose career is more typical of the life, say, of the average European peasant woman, or of the women of barbarous tribes, its grace and youthfulness pass so quickly into stoutness, obesity, and withered old age. How positively girlish and taking is the daisy during the first few days of its blooming, while its snow-white rays yet stand straight up and shield its tender centre somewhat as a hood shields a girl's face! Presently it becomes a perfect disk and bares its face to the sun; this is the stage of its young womanhood. Then its yellow centre — its body — begins to swell and become gross, the rays slowly turn brown, and finally wither up and drop. It is a flower no longer, but a receptacle packed with ripening seeds.

A relative of the daisy, the orange-colored hawk-weed (*Hieracium aurantiacum*), which within the past twenty years has spread far and wide over New York and New England, is often at the height of its beauty in August, when its deep vivid orange is a delight to the eye. It repeats in our meadows and upon our hilltops the flame of the columbine of May, intensified. The personified August with these flowers in her hair would challenge our admiration and not our criticism. Unlike the daisy, it quickly sprouts again when cut down with the grass in the meadows, and renews its bloom. Parts of New England, at least, have a native August flower quite as brilliant as the hawkweed just described, and far less a usurper; I refer to meadow-beauty, or rhexia, found near the coast, which suggests a purple evening primrose.

Nature has, for the most part, lost her delicate tints in August. She is tanned, hirsute, freckled, like one long exposed to the sun. Her touch is strong and vivid. The coarser, commoner wayside flowers now appear, — vervain, eupatorium, mimulus, the various mints, asters, golden-rod, thistles, fireweed, mulleins, motherwort, catnip, blueweed, turtle-head, sunflowers, clematis, evening primrose, lobe-lia, gerardia, and, in the marshes of the lower Hudson, marshmallows, and vast masses of the purple loosestrife. Mass and intensity take the place of delicacy and furtiveness. The spirit of Nature has grown bold and aggressive; it is rank and coarse; she flaunts her weeds in our faces. She wears a thistle on her bosom. But I must not forget the delicate rose gerardia, which she also wears upon her bosom, and which suggests that, before the season closes, Nature is getting her hand ready for her delicate spring flora. With me this gerardia lines open paths over dry knolls in the woods, and its little purple bells and smooth, slender leaves form one of the most exquisite tangles of flowers and foliage of the whole summer. It is August matching the color and delicacy of form of the fringed polygala of May. I know a half-wild field bordering a wood, which is red with strawberries in June and pink with gerardia in August.

One may still gather the matchless white pond-lily in this month, though this flower is in the height of its glory earlier in the season, except in the northern lakes.

A very delicate and beautiful marsh flower that may be found on the borders of lakes in northern New York and New England is the horned bladderwort, — yellow, fragrant, and striking in form, like a miniature old-fashioned bonnet, when bonnets covered the head and projected beyond the face, instead of hovering doubtfully above the scalp. The horn curves down and out like a long chin from a face hidden within the bonnet. I have found this rare flower in the Adirondacks and in Maine. It may doubtless be found in Canada, and in Michigan and Wisconsin. Britton and Brown say "south to Florida and Texas." It is the most fragrant August flower known to me. This month has not many fragrant flowers to boast of. Besides the above and the pond-lily I recall two others, — the small purple fringed-orchis and a species of lady's-tresses (*Spiranthes cernua*).

The characteristic odors of August are from fruit — grapes, peaches, apples, pears, melons — and the ripening grain; yes, and the blooming buckwheat. Of all the crop and farm odors this last is the most pronounced and honeyed, rivaling that of the flower-ing locust of May and of the linden in July.

The mistakes of our lesser poets in dealing with nature themes might furnish me with many a text in this connection. Thus one of them makes the call of the phœbe-bird prominent in August. One would infer from the poem that the phœbe was not heard during any other month. Now it is possible that the poet heard the phœbe in August, but if so, the occurrence was exceptional, and it is more proba-ble that it was the wood pewee that he heard. The phœbe is most noticeable in April and early May, and its characteristic call is not often heard till the sun is well up in the sky. Most of our song-birds are silent in August, or sing only fitfully, as do the

song sparrow and the oriole. The real August songster, and the bird that one comes to associate with
the slow, drowsy days, is the indigo-bird. After
midsummer its song, delivered from the top of some
small tree in the pasture or a bushy field, falls upon
the ear with a peculiar languid, midsummery effect.
The boys and girls gathering raspberries and blackberries hear it; the stroller through the upland fields,
or lounger in the shade of maple or linden, probably
hears no bird-song but this, if he even distinguishes
this from the more strident insect voices. The plumage of the bird is more or less faded by this time,
the vivid indigo of early June is lightly brushed with
a dull sooty shade, but the song is nearly as full as
the earlier strain, and in the dearth of bird voices is
even more noticeable. I do not now recall that any
of our poets have embalmed this little cerulean songster in their verse.

One may also occasionally hear the red-eyed vireo
in August, but it is low tide with him too. His song
has a reminiscent air, like that of the indigo. The
whip-poor-will calls fitfully in this month, and may
be heard even in September; but he quickly checks
himself, as if he knew it was out of season. In the
Adirondacks I have heard the speckled Canada
warbler in August, and the white-throated sparrow.
But nearly all the migratory birds begin to get restless during this month. They cut loose from their
nesting-haunts and drift through the woods in pro-

miscuous bands, and many of them start on their southern journey. From my woods along the Hudson the warblers all disappear before the middle of the month. Some of them are probably in hiding during the moulting season. The orioles begin to move south about the middle of the month, and by the first of September the last of them have passed. They occasionally sing in a suppressed tone during this migration, probably the young males trying their instruments. It is at this time, when full of frolic and mischief, like any other emigrants with faces set to new lands, that they make such havoc in the Hudson River vineyards. They seem to puncture the grapes in the spirit of pure wantonness, or as if on a wager as to who can puncture the most. The swallows — the cliff and the barn — all leave in August, usually by the 20th, though the swift may be seen as late as October. I notice that our poets often detain the swallows much beyond the proper date. One makes them perch upon the barn in October. Another makes them noisy about the eaves in Indian summer. An English poet makes the swallow go at November's bidding. The tree swallow may often be seen migrating in countless numbers along the coast in early October, but long ere this date the barn and the cliff swallows are in tropical climes. They begin to flock, and apparently rehearse the migrating programme, in July.

The bobolinks go in early August with the red-

shouldered starlings, and along the Potomac and
Chesapeake Bay become the reed-birds of sportsmen. One often hears them in this month calling
from high in the air as they journey southward from
more northern latitudes.

About the most noticeable bird of August in New
York and New England is the yellowbird, or goldfinch. This is one of the last birds to nest, seldom
hatching its eggs till late in July. It seems as if a
particular kind of food were required to rear its
brood, which cannot be had at an earlier date. The
seed of the common thistle is apparently its mainstay. There is no prettier sight at this season than
a troop of young goldfinches, led by their parents,
going from thistle to thistle along the roadside and
pecking the ripe heads to pieces for the seed. The
plaintive call of the young is one of the characteristic August sounds. Their nests are frequently
destroyed, or the eggs thrown from them, by the
terrific July thunder-showers. Last season a pair
had a nest on the slender branch of a maple in front
of the door of the house where I was staying. The
eggs were being deposited, and the happy pair had
many a loving conversation about them many times
each day, when one afternoon a very violent storm
arose which made the branches of the trees stream
out like wildly disheveled hair, quite turning over
those on the windward side, and emptying the pretty
nest of its eggs. In such cases the birds build anew.

— a delay that may bring the incubation into August. Such an accident had probably befallen a pair of which I one season made this note in my note-book, under date of August 6: —

"A goldfinches' nest in the maple-tree near the window where I write, the female sitting on four pale bluish-white eggs; the male feeds her on the nest; whenever she hears his voice she calls incessantly, much after the manner of the young birds, — the only case I recall of the sitting bird calling while in the act of incubation. The male evidently brings the food in his crop, or at least well back in his beak or throat, as it takes him several moments to deliver it to his mate, which he does by separate morsels. The male, when disturbed by a rival, utters the same note as he pursues his enemy from point to point that the female does when calling to him. It does not sound like a note of anger, but of love and con-fidence."

As the bird-songs fail, the insect harpers and fid-dlers begin. August is the heyday of these musi-cians. The katydid begins to "work her chromatic reed" early in the month, and with her comes that pulsing, purring monotone of the little pale tree-crickets. These last fill the August twilight with a soft, rhythmic undertone of sound, which forms a sort of background for the loud, strident notes of the katydids.

August, too, is the month of the screaming, high-sailing hawks. The young are now fully fledged, and they love to circle and scream far above the mountain's crest all the tranquil afternoon. Sometimes one sees them against the slowly changing and swelling thunder-heads that so often burden the horizon at this season.

It is on the dewy August mornings that one notices the webs of the little spiders in the newly mown meadows. They look like gossamer napkins spread out upon the grass, — thousands of napkins far and near. The farmer looks upon it as a sign of rain; but the napkins are there every day; only a heavier dew makes them more pronounced one morning than another.

Along the paths where my walks oftenest lead me in August, in rather low, bushy, wet grounds, the banner flower is a species of purple boneset, or trumpet-weed, so called, I suppose, because its stem is hollow. It often stands up seven or eight feet high, crowned with a great mass of dull purple bloom, and leads the ranks of lesser weeds and plants like a great chieftain. Its humbler servitors are white boneset and swamp milkweed, while climbing boneset trails its wreaths over the brookside bushes not far away. A much more choice and brilliant purple, like some invasion of metropolitan fashion into a rural congregation, is given to a near-by marsh by the purple loosestrife. During the latter half of August the bog is all aflame with it. There is a wonderful style about this plant, either singly or in masses. Its suggestion is as distinctly feminine as that of the trumpet-weed is masculine.

When the poet personifies August, let him fill her arms with some of these flowers, or place upon her brows a spray of wild clematis, which during this month throws its bridal wreaths so freely over our bushy, unkempt waysides and fence corners. After you have crowned and adored your personified August in this way, then give the finishing touch with the scarlet raceme of the cardinal flower, flaming from the sheaf of ranker growths in her arms. How this brilliant bit of color, glassing itself in a dark, still pool, lights up and affects the vague, shadowy background upon which it is projected!

In August the "waters blossom." This is the term the country people in my section apply to a phenomenon which appears in the more sluggish streams and ponds during this month. When examined closely, the water appears to be filled with particles of very fine meal. I suspect, though I do not know, that these floating particles are the spores of some species of fresh-water alga; or they may be what are called zoöspores. The algæ are at their rankest during August. Great masses of some species commonly called "frogs' spawn" rise to the surface of the Hudson and float up and down with the tide, — green unclean-looking masses, many yards in extent. The dog-star seems to invoke these fermenting masses from the deep. They suggest decay, but they are only the riot of the lower forms of vegetable life.

August, too, is the month of the mushrooms, — those curious abnormal flowers of a hidden or subterranean vegetation, invoked by heat and moisture from darkness and decay as the summer wanes. Do they not suggest something sickly and uncanny in Nature? her unwholesome dreams and night fancies, her pale superstitions ; her myths and legends and occult lore taking shape in them, spectral and fantastic, at times hinting something libidinous and unseemly: vegetables with gills, fibreless, bloodless; earth-flesh, often offensive, unclean, immodest, often of rare beauty and delicacy, of many shades and colors — creamy white, red, yellow, brown, — now the hue of an orange, now of a tomato, now of a potato, some edible, some poisonous, some shaped like spread umbrellas, some like umbrellas reversed by the wind, — the sickly whims and fancies of Nature, some imp of the earth mocking and travestying the things of the day. Under my evergreens I saw a large white disk struggling up through the leaves and the débris like the full moon through clouds and vapors. This simile is doubtless suggested to my mind by a line of a Southern poet, Madison Cawein, which I look upon as one of the best descriptive lines in recent nature poetry: —

FAR AND NEAR

" The slow toadstool comes bulging, moony white,
   Through loosening loam.''

Sometimes this moon of the loam is red, or golden,
or bronzed; or it is so small that it suggests only a
star. The shy wood folk seem to know the edible
mushrooms, and I notice often eat away the stalk
and nibble at the top or pileus.

One day two friends came to see me with something wrapped up in their handkerchiefs. They said
they had brought their dinner with them, — they
had gathered it in the woods as they came along,
— beefsteak mushrooms. The beefsteak was duly
cooked and my friends ate of it with a relish. A por-
tion was left, which my dog attacked rather doubtingly, and then turned away from, with the look of
one who has been cheated. Mock-meat, that is what
it was, — a curious parody upon a steak, as the dog
soon found out. I know a man who boasts of having identified and eaten seventy-five different species. When the season is a good one for mushrooms,
he snaps his fingers at the meat trust, even going to
the extent of drying certain kinds to be used for
soup in the winter.

The decay of a mushroom parodies that of real
flesh, — a kind of unholy rotting ending in blackness and stench. Some species imitate jelly, — mock
calves'-foot jelly, which soon melts down and becomes an uncanny mass. Occasionally I see a blue-

gilled mushroom, — an infusion of indigo in its cells. How forbidding it looks! Yesterday in the August woods I saw a tiny mushroom like a fairy parasol of a Japanese type, — its top delicately fluted.

During the steaming, dripping, murky, and muggy dog-days that sometimes come the latter half of August, how this fungus growth runs riot in the woods and in the fields too, — a kind of sacrilegious vegetation mocking Nature's saner and more wholesome handiwork, — the flowers of death, vegetable spectres.

August days are for the most part tranquil days; the fret and hurry of the season are over. We are on the threshold of autumn. Nature dreams and meditates; her veins no longer thrill with the eager, frenzied sap; she ripens and hardens her growths; she concentrates; she begins to make ready for winter. The buds for next year are formed during this month, and her nuts and seeds and bulbs finish storing up food for the future plant.

From my outlook upon the Hudson the days are placid, the river is placid, the boughs of the trees gently wag, the bees make vanishing-lines through the air. The passing boats create a great commotion in the water, converting it from a cool, smooth, shadowy surface to one pulsing and agitated. The pulsations go shoreward in long, dark, rolling, glassy swells. The grapes are purpling in the vineyard; the apples and pears are coloring in the orchard;

the corn is glazing in the field; the oats are ripe for the cradle; grasshoppers poise and shuffle above the dry road; yellow butterflies mount upward face to face; thistledown drifts by on the breeze; a sparrow sings fitfully now and then; dusty wheelmen go by on their summer vacation tours; boats appear upon the river loaded with gay excursionists, and on every hand the stress and urge of life have abated.

ANIMAL COMMUNICATION

Feeding a starving bird
during a trip to Georgia

EDITOR'S NOTE

"Animal Communication" is one of several essays on animal behavior that Burroughs wrote during the first decade of the century. Originally called "On Humanizing the Animals," the essay first appeared in *Century* in March 1904. As the original title indicates, the essay rose out of the "nature faker" controversy that began with Burroughs's attack on sentimental nature writers in "Real and Sham Natural History," published in the *Atlantic* in March 1903. When Burroughs collected several of these essays into *Ways of Nature* (1905), he added footnotes and cross references to give the book a thematic coherence. Burroughs drew on the authority of secondary sources and the results of scientific experimentation to substantiate his own observations about animal life and instinct. He regularly condemned those who attributed conceptual thought or speculative reason to animals while he simultaneously advocated similarities between human and animal emotion, memory, and perception. Nothing is more controversial in nature writing than the comparisons naturalists make between human beings and other animals. Burroughs himself has been attacked for violating the strictures against anthropomorphism he delineates in this essay. At the same time he has been attacked for being too mechanistic in his distinction between animal instinct and human consciousness. Whether consciousness is an evolved ability shared by humans with other animals forms a continuing debate which can profit from Burroughs's observation that most misconceptions in natural history arise from a careless use of words.

F.B.

ANIMAL COMMUNICATION

THE notion that animals consciously train and educate their young has been held only tentatively by European writers on natural history. Darwin does not seem to have been of this opinion at all. Wallace shared it at one time in regard to the birds, — their songs and nest-building, — but abandoned it later, and fell back upon instinct or inherited habit. Some of the German writers, such as Brehm, Büchner, and the Müllers, seem to have held to the notion more decidedly. But Professor Groos had not yet opened their eyes to the significance of the play of animals. The writers mentioned undoubtedly read the instinctive play of animals as an attempt on the part of the parents to teach their young.

That the examples of the parents in many ways stimulate the imitative instincts of the young is quite certain, but that the parents in any sense aim at instruction is an idea no longer held by writers on animal psychology.

Of course it all depends upon what we mean by teaching. Do we mean the communication of knowledge, or the communication of emotion? It seems to me that by teaching we mean the former. Man alone communicates knowledge; the lower animals communicate feeling or emotion. Hence their communications always refer to the present, never to the past or to the future.

That birds and beasts do communicate with each other, who can doubt? But that they impart knowledge, that they have any knowledge to impart, in the strict meaning of the word, any store of ideas or mental concepts — that is quite another matter. Teaching implies such store of ideas and power to impart them. The subconscious self rules in the animal; the conscious self rules in man, and the conscious self alone can teach or communicate knowledge. It seems to me that the cases of the deer and the antelope, referred to by President Roosevelt in the letter to me quoted in the last chapter, show the communication of emotion only.

Teaching implies reflection and judgment; it implies a thought of, and solicitude for, the future. "The young will need this knowledge," says the human parent, "and so we will impart it to them now." But the animal parent has consciously no knowledge to impart, only fear or suspicion. One may affirm almost anything of trained dogs and of dogs generally. I can well believe that the setter bitch spoken of by the President punished her pup when it flushed a bird, — she had been punished herself for the same offense, — but that the act was expressive of anything more than her present anger, that she was in any sense trying to train and instruct her pup, there is no proof.

But with animals that have not been to school to man, all ideas of teaching must be rudimentary indeed. How could a fox or a wolf instruct its young in such matters as traps? Only in the presence of the trap, certainly; and then the fear of the trap would be communicated to the young through natural instinct. Fear, like joy or curiosity, is contagious among beasts and birds, as it is among men; the young fox or wolf would instantly share the emotion of its parent in the presence of a trap. It is very important to the wild creatures that they have a quick apprehension of danger, and as a matter of fact they have. One wild and suspicious duck in a flock will often defeat the best laid plans of the duck-hunter. Its suspicions are quickly communicated to all its fellows: not through any conscious effort on its part to do so, but through the law of natural contagion above referred to. Where any bird or beast is much hunted, fear seems to be in the air, and their fellows come to be conscious of the danger which they have not experienced.

What an animal lacks in wit it makes up in caution. Fear is a good thing for the wild creatures to have in superabundance. It often saves them from real danger. But how undiscriminating it is! It is said that an iron hoop or wagon-tire placed around a setting hen in the woods will protect her from the foxes.

Animals are afraid on general principles. Anything new and strange excites their suspicions. In a herd of animals, cattle, or horses, fear quickly becomes a panic and rages like a conflagration. Cattlemen in the West found that any little thing at night might kindle the spark in their herds and sweep the whole mass away in a furious stampede. Each animal excites every other, and the multiplied fear of the herd is something terrible. Panics among men are not much different.

In a discussion like the present one, let us use words in their strict logical sense, if possible. Most of the current misconceptions in natural history, as in other matters, arise from a loose and careless use of words. One says teach and train and instruct, when the facts point to instinctive imitation or unconscious communication.

That the young of all kinds thrive better and develop more rapidly under the care of their parents than when deprived of that care is obvious enough. It would be strange if it were not so. Nothing can quite fill the place of the mother with either man or bird or beast. The mother provides and protects. The young quickly learn of her through the natural instinct of imitation. They share her fears, they follow in her footsteps, they look to her for protection;

it is the order of nature. They are not trained in the way they should go, as a child is by its human parents — they are not trained at all; but their natural instincts doubtless act more promptly and surely with the mother than without her. That a young kingfisher or a young osprey would, in due time, dive for fish, or a young marsh hawk catch mice and birds, or a young fox or wolf or coon hunt for its proper prey without the parental example, admits of no doubt at all; but they would each probably do this thing earlier and better in the order of nature than if that order were interfered with.

The other day I saw a yellow-bellied woodpecker alight upon a decaying beech and proceed to drill for a grub. Two of its fully grown young followed it and, alighting near, sidled up to where the parent was drilling. A hasty observer would say that the parent was giving its young a lesson in grub-hunting, but I read the incident differently. The parent bird had no thought of its young. It made passes at them when they came too near, and drove them away. Presently it left the tree, whereupon one of the young examined the hole its parent had made and drilled a little on its own account. A parental example like this may stimulate the young to hunt for grubs earlier than they would otherwise do, but this is merely conjecture. There is no proof of it, nor can there be any.

The mother bird or beast does not have to be

instructed in her maternal duties: they are instinctive with her; it is of vital importance to the continuance of the species that they should be. If it were a matter of instruction or acquired knowledge, how precarious it would be!

The idea of teaching is an advanced idea, and can come only to a being that is capable of returning upon itself in thought, and that can form abstract conceptions — conceptions that float free, so to speak, dissociated from particular concrete objects.

If a fox, or a wolf, for instance, were capable of reflection and of dwelling upon the future and upon the past, it might feel the need of instructing its young in the matter of traps and hounds, if such a thing were possible without language. When the cat brings her kitten a live mouse, she is not thinking about instructing it in the art of dealing with mice, but is intent solely upon feeding her young. The kitten already knows, through inheritance, about mice. So when the hen leads her brood forth and scratches for them, she has but one purpose — to provide them with food. If she is confined to the coop, the chickens go forth and soon scratch for themselves and snap up the proper insect food.

The mother's care and protection count for much, but they do not take the place of inherited instinct. It has been found that newly hatched chickens, when left to themselves, do not know the difference between edible and non-edible insects, but that they soon learn. In such matters the mother hen, no doubt, guides them.

A writer in "Forest and Stream," who has since published a book about his "wild friends," pushes this notion that animals train their young so far that it becomes grotesque. Here are some of the things that this keen observer and exposer of "false natural history" reports that he has seen about his cabin in the woods: He has seen an old crow that hurriedly flew away from his cabin door on his sudden appearance, return and beat its young because they did not follow quickly enough. He has seen a male chewink, while its mate was rearing a second brood, take the first brood and lead them away to a bird-resort (he probably meant to say to a bird-nursery or kindergarten); and when one of the birds wandered back to take one more view of the scenes of its infancy, he has seen the father bird pounce upon it and give it a "severe whipping and take it to the resort again."

He has seen swallows teach their young to fly by gathering them upon fences and telegraph wires and then, at intervals (and at the word of command, I suppose), launching out in the air with them, and swooping and circling about. He has seen a song sparrow, that came to his dooryard for fourteen years (he omitted to say that he had branded him and so knew his bird), teach *his year-old boy to sing* (the italics are mine). This hermit-inclined sparrow wanted to " desert the fields for a life in the woods," but his "wife would not consent." Many a feather-less biped has had the same experience with his society-spoiled wife. The puzzle is, how did this masterly observer know that this state of affairs existed between this couple? Did the wife tell him, or the husband? "Hermit" often takes his visitors to a wood thrushes' singing-school, where, "as the birds forget their lesson, they drop out one by one."

He has seen an old rooster teaching a young rooster to crow! At first the old rooster crows mostly in the morning, but later in the season he crows throughout the day, at short intervals, to show the young "the proper thing." "Young birds re-moved out of hearing will not learn to crow." He hears the old grouse teaching the young to drum in the fall, though he neglects to tell us that he has seen the young in attendance upon these lessons. He has seen a mother song sparrow helping her two-year-old daughter build her nest. He has discov-ered that the cat talks to her kittens with her ears: when she points them forward, that means "yes;" when she points them backward, that means "no." Hence she can tell them whether the wagon they hear approaching is the butcher's cart or not, and thus save them the trouble of looking out.

And so on through a long list of wild and domes-tic creatures. At first I suspected this writer was covertly ridiculing a certain other extravagant "ob-server," but a careful reading of his letter shows him

to be seriously engaged in the worthy task of exposing "false natural history."

Now the singing of birds, the crowing of cocks,
the drumming of grouse, are secondary sexual characteristics. They are not necessary to the lives of the
creatures, and are probably more influenced by imitation than are the more important instincts of self-
preservation and reproduction. Yet the testimony is
overwhelming that birds will sing and roosters crow
and turkeys gobble, though they have never heard
these sounds; and, no doubt, the grouse and the
woodpeckers drum from promptings of the same
sexual instinct.

I do not wish to accuse "Hermit" of willfully per-
verting the facts of natural history. He is one of
those persons who read their own fancies into whatever they look upon. He is incapable of disinterested
observation, which means he is incapable of observation at all in the true sense. There are no animals
that signal to each other with their ears. The movements of the ears follow the movements of the eye.
When an animal's attention is directed to any object or sound, its ears point forward; when its attention is relaxed, the ears fall. But with the cat tribe
the ears are habitually erect, as those of the horse
are usually relaxed. They depress them and revert
them, as do many other animals, when angered or
afraid.

Certain things in animal life lead me to suspect

that animals have some means of communication with one another, especially the gregarious animals, that is quite independent of what we mean by language. It is like an interchange or blending of subconscious states, and may be analogous to telepathy among human beings. Observe what a unit a flock of birds becomes when performing their evolutions in the air. They are not many, but one, turning and flashing in the sun with a unity and a precision that it would be hard to imitate. One may see a flock of shore-birds that behave as one body: now they turn to the sun a sheet of silver; then, as their dark backs are presented to the beholder, they almost disappear against the shore or the clouds. It would seem as if they shared in a communal mind or spirit, and that what one felt they all felt at the same instant.

In Florida I many times saw large schools of mullets fretting and breaking the surface of the water with what seemed to be the tips of their tails. A large area would be agitated and rippled by the backs or tails of a host of fishes. Then suddenly, while I looked, there would be one splash and every fish would dive. It was a multitude, again, acting as one body. Hundreds, thousands of tails slapped the water at the same instant and were gone.

When the passenger pigeons were numbered by millions, the enormous clans used to migrate from one part of the continent to another. I saw the last flight of them up the Hudson River valley in the spring of 1875. All day they streamed across the sky. One purpose seemed to animate every flock and every bird. It was as if all had orders to move to the same point. The pigeons came only when there was beech-mast in the woods. How did they know we had had a beech-nut year? It is true that a few straggling bands were usually seen some days in advance of the blue myriads: were these the scouts, and did they return with the news of the beech-nuts? If so, how did they communicate the intelligence and set the whole mighty army in motion?

The migrations among the four-footed animals that sometimes occur over a large part of the coun-try — among the rats, the gray squirrels, the rein-deer of the north — seem to be of a similar char-acter. How does every individual come to share in the common purpose? An army of men attempting to move without leaders and without a written or spoken language becomes a disorganized mob. Not so the animals. There seems to be a community of mind among them in a sense that there is not among men. The pressure of great danger seems to develop in a degree this community of mind and feeling among men. Under strong excitement we revert more or less to the animal state, and are ruled by instinct. It may well be that telepathy — the power to project one's mental or emotional state so as to impress a friend at a distance — is a power which we have carried over from our remote animal ancestors. However this may be, it is certain that the sensitiveness of birds and quadrupeds to the condition of one another, their sense of a common danger, of food supplies, of the direction of home under all circumstances, point to the possession of a power which is only rudimentary in us.

Some observers explain these things on the theory that the flocks of birds have leaders, and that their surprising evolutions are guided by calls or signals from these leaders, too quick or too fine for our eyes or ears to catch. I suppose they would explain the movements of the schools of fish and the simultaneous movements of a large number of land animals on the same theory. I cannot accept this explanation. It is harder for me to believe that a flock of birds has a code of calls or signals for all its evolutions — now right, now left, now mount, now swoop — which each individual understands on the instant, or that the hosts of the wild pigeons had their captains and signals, than to believe that out of the flocking instinct there has grown some other instinct or faculty, less understood, but equally potent, that puts all the members of a flock in such complete rapport with one another that the purpose and the desire of one become the purpose and the desire of all. There is nothing in this state of things analogous to a military organization. The relation among the members of the flock is rather that of creatures sharing

spontaneously the same subconscious or psychic
state, and acted upon by the same hidden influence,
in a way and to a degree that never occur among
men.

The faculty or power by which animals find the
way home over or across long stretches of country
is quite as mysterious and incomprehensible to us
as the spirit of the flock to which I refer. A hive
of bees evidently has a collective purpose and plan
that does not emanate from any single individual
or group of individuals, and which is understood
by all without outward communication.

Is there anything which, without great violence
to language, may be called a school of the woods?
In the sense in which a playground is a school — a
playground without rules or methods or a director
— there is a school of the woods. It is an unkept, an
unconscious school or gymnasium, and is entirely
instinctive. In play the young of all animals, no
doubt, get a certain amount of training and disci-
plining that helps fit them for their future careers;
but this school is not presided over or directed by
parents, though they sometimes take part in it. It is
spontaneous and haphazard, without rule or system;
but is, in every case, along the line of the future
struggle for life of the particular bird or animal.
A young marsh hawk which we reared used to play
at striking leaves or bits of bark with its talons;
kittens play with a ball, or a cob, or a stick, as if

it were a mouse; dogs race and wrestle with one
another as in the chase; ducks dive and sport in the
water; doves circle and dive in the air as if escaping from a hawk; birds pursue and dodge one
another in the same way; bears wrestle and box;
chickens have mimic battles; colts run and leap;
fawns probably do the same thing; squirrels play
something like a game of tag in the trees; lambs butt
one another and skip about the rocks; and so on.

In fact, nearly all play, including much of that of
man, takes the form of mock battle, and is to that
extent an education for the future. Among the carnivora it takes also the form of the chase. Its spring
and motive are, of course, pleasure, and not education; and herein again is revealed the cunning of
nature — the power that conceals purposes of its
own in our most thoughtless acts. The cat and the
kitten play with the live mouse, not to indulge the
sense of cruelty, as some have supposed, but to indulge in the pleasure of the chase and unconsciously
to practice the feat of capture. The cat rarely plays
with a live bird, because the recapture would be more
difficult, and might fail. What fisherman would not
like to take his big fish over and over again, if he
could be sure of doing it, not from cruelty, but for
the pleasure of practicing his art? For further light
on the subject of the significance of the play of animals, I refer the reader to the work of Professor
Karl Groos called "The Play of Animals."

ANIMAL COMMUNICATION

One of my critics has accused me of measuring all things by the standard of my little farm — of thinking that what is not true of animal life there is not true anywhere. Unfortunately my farm *is* small — hardly a score of acres — and its animal life very limited. I have never seen even a porcupine upon it; but I have a hill where one might roll down, should one ever come my way and be in the mood for that kind of play. I have a few possums, a woodchuck or two, an occasional skunk, some red squirrels and rabbits, and many kinds of song-birds. Foxes occasionally cross my acres; and once, at least, I saw a bald eagle devouring a fish in one of my apple-trees. Wild ducks, geese, and swans in spring and fall pass across the sky above me. Quail and grouse invade my premises, and of crows I have, at least in bird-nesting time, too many.

But I have a few times climbed over my pasture wall and wandered into distant fields. Once upon a time I was a traveler in Asia for the space of two hours — an experience that ought to have yielded me some startling discoveries, but did not. Indeed, the wider I have traveled and observed nature, the more I am convinced that the wild creatures behave just about the same in all parts of the country; that is, under similar conditions. What one observes truly about bird or beast upon his farm of ten acres, he will not have to unlearn, travel as wide or as far as he will. Where the animals are much hunted, they are of course much wilder and more cunning than where they are not hunted. In the Yellowstone National Park we found the elk, deer, and mountain sheep singularly tame; and in the summer, so we were told, the bears board at the big hotels. The wild geese and ducks, too, were tame; and the red-tailed hawk built its nest in a large dead oak that stood quite alone near the side of the road. With us the same hawk hides its nest in a tree in the dense woods, because the farmers unwisely hunt and destroy it. But the cougars and coyotes and bobcats were no tamer in the park than they are in other places where they are hunted.

Indeed, if I had elk and deer and caribou and moose and bears and wildcats and beavers and otters and porcupines on my farm, I should expect them to behave just as they do in other parts of the country under like conditions: they would be tame and docile if I did not molest them, and wild and fierce if I did. They would do nothing out of character in either case.

Your natural history knowledge of the East will avail you in the West. There is no country, says Emerson, in which they do not wash the pans and spank the babies; and there is no country where a dog is not a dog, or a fox a fox, or where a hare is ferocious, or a wolf lamblike. The porcupine behaves in the Rockies just as he does in the Catskills;

the deer and the moose and the black bear and the beaver of the Pacific slope are almost identical in their habits and traits with those of the Atlantic slope.

In my observations of the birds of the far West, I went wrong in my reckoning but once: the Western meadowlark has a new song. How or where he got it is a mystery; it seems to be in some way the gift of those great, smooth, flowery, treeless, dimpled hills. But the swallow was familiar, and the robin and the wren and the highhole, while the woodchuck I saw and heard in Wyoming might have been the "chuck" of my native hills. The eagle is an eagle the world over. When I was a boy I saw, one autumn day, an eagle descend with extended talons upon the backs of a herd of young cattle that were accompanied by a cosset-sheep and were feeding upon a high hill. The object of the eagle seemed to be to separate the one sheep from the cattle, or to frighten them all into breaking their necks in trying to escape him. But neither result did he achieve. In the Yellowstone Park, President Roosevelt and Major Pitcher saw a golden eagle trying the same tactics upon a herd of elk that contained one yearling. The eagle doubtless had his eye upon the yearling, though he would probably have been quite satisfied to have driven one of the older ones down a precipice. His chances of a dinner would have been equally good.

WAYS OF NATURE

There is one particular in which the bird families are much more human than our four-footed kindred. I refer to the practice of courtship. The male of all birds, so far as I know, pays suit to the female and seeks to please and attract her.[1] This the quadrupeds do not do; there is no period of courtship among them, and no mating or pairing as among the birds. The male fights for the female, but he does not seek to win her by delicate attentions. If there are any exceptions to this rule, I do not know them. There seems to be among the birds something that is like what is called romantic love. The choice of mate seems always to rest with the female,[1] while among the mammals the female shows no preference at all.

Among our own birds, the prettiest thing I know of attending the period of courtship, or preliminary to the match-making, is the spring musical festival and reunion of the goldfinches, which often lasts for days, through rain and shine. In April or May, apparently all the goldfinches from a large area collect in the top of an elm or a maple and unite in a prolonged musical festival. Is it a contest among the males for the favor of the females, or is it the spontaneous expression of the gladness of the whole clan at the return of the season of life and love? The birds seem to pair soon after, and doubtless the concert of voices has some reference to that event.

[1] Except in the case of certain birds of India and Australia.

ANIMAL COMMUNICATION

There is one other human practice often attributed to the lower animals that I must briefly consider, and that is the practice, under certain circumstances, of poisoning their young. One often
hears of caged young birds being fed by their parents
for a few days and then poisoned; or of a mother
fox poisoning her captive young when she finds that
she cannot liberate him; and such stories obtain
ready credence with the public, especially with the
young. To make these stories credible, one must
suppose a school of pharmacy, too, in the woods.

"The worst thing about these poisoning stories,"
writes a friend of mine, himself a writer of nature-
books, "is the implied appreciation of the full effect
and object of poison — the comprehension by the
fox, for instance, that the poisoned meat she may
be supposed to find was placed there for the object
of killing herself (or some other fox), and that she
may apply it to another animal for that purpose.
Furthermore, that she understands the nature of
death — that it brings 'surcease of sorrow,' and
that death is better than captivity for her young one.
How did she acquire all this knowledge? Where
was her experience of its supposed truth obtained?
How could she make so fine and far-seeing a judgment, wholly out of the range of brute affairs, and
so purely philosophical and humanly ethical? It
violates every instinct and canon of natural law,
which is for the preservation of life at all hazards.

This is simply the human idea of 'murder.' Animals kill one another for food, or in rivalry, or in blind ferocity of predatory disposition; but there is not a particle of evidence that they 'commit murder' for ulterior ends. It is questionable whether they comprehend the condition called death, or its nature, in any proper sense."

On another occasion I laughed at a recent nature writer for his credulity in half-believing the story told him by a fisherman, that the fox catches crabs by using his tail as a bait; and yet I read in Romanes that Olaus, in his account of Norway, says he has seen a fox do this very thing among the rocks on the sea-coast.[1] One would like to cross-question Olaus before accepting such a statement. One would as soon expect a fox to put his brush in the fire as in the water. When it becomes wet and bedraggled, he is greatly handicapped as to speed. There is no doubt that rats will put their tails into jars that contain liquid food they want, and then lick them off, as Romanes proved; but the rat's tail is not a brush, nor in any sense an ornament. Think what the fox-and-crab story implies! Now the fox is entirely a land animal, and lives by preying upon land crea-

---

[1] A book published in London in 1783, entitled *A Geographical, Historical, and Commercial Grammar and the Present State of the Several Kingdoms of the World*, among other astonishing natural history notes, makes this statement about the white and red fox of Norway: " They have a particular way of drawing crabs ashore by dipping their tails in the water, which the crab lays hold of."

tures, which it follows by scent or sight. It can neither see nor smell crabs in the deep water, where crabs are usually found. How should it know that there are such things as crabs? How should it know that they can be taken with bait and line or by fishing for them? When and how did it get this experience? This knowledge belongs to man alone. It comes through a process of reasoning that he alone is capable of. Man alone of land animals sets traps and fishes. There is a fish called the angler (*Lophius piscatorius*), which, it is said on doubtful authority, by means of some sort of appendages on its head angles for small fish; but no competent observer has reported any land animal doing so. Again, would a crab lay hold of a mass of fur like a fox's tail? — even if the tail could be thrust deep enough into the water, which is impossible. Crabs, when not caught with hand-nets, are usually taken in water eight or ten feet deep. They are baited and caught with a piece of meat tied to a string, but cannot be lifted to the surface till they are eating the meat, and then a dip-net is required to secure them. The story, on the whole, is one of the most preposterous that ever gained credence in natural history.

Good observers are probably about as rare as good poets. Accurate seeing, — an eye that takes in the whole truth, and nothing but the truth, — how rare indeed it is! So few persons know or can tell exactly what they see; so few persons can draw a right inference from an observed fact; so few persons can keep from reading their own thoughts and preconceptions into what they see ; only a person with the scientific habit of mind can be trusted to report things as they are. Most of us, in observing the wild life about us, see more or see less than the truth. We see less when our minds are dull, or preoccupied, or blunted by want of interest. This is true of most country people. We see more when we read the lives of the wild creatures about us in the light of our human experience, and impute to the birds and beasts human motives and methods. This is too often true of the eager city man or woman who sallies out into the country to study nature.

The tendency to sentimentalize nature has, in our time, largely taken the place of the old tendency to demonize and spiritize it. It is anthropomorphism in another form, less fraught with evil to us, but equally in the way of a clear understanding of the life about us.

THE GRIST OF THE GODS

Studying geology atop his boyhood rock
at his Catskill family farm

EDITOR'S NOTE

As early as 1878 Burroughs contemplated writing an essay about "dirt," but he did not fulfill his plan until thirty years later when he published "The Divine Soil" and "The Grist of the Gods" (1908). Stimulated by his trips out west and his correspondence with John Muir and Bailey Willis of the United States Geological Survey, Burroughs wrote numerous essays about geology during the last decades of his life. In 1912 he published essays on Yosemite, the Grand Canyon, and the Catskills in *Time and Change*, a book devoted primarily to geology, and in 1916 he published another long essay on his native hills called "The Friendly Rocks" in *Under the Apple Trees*. In "The Grist of the Gods," reprinted here from *Leaf and Tendril* (1908), Burroughs draws from his reading in astronomy, physics, and chemistry as well as geology. In one of his most visionary essays he presents the earth as a living entity covered by a thin film of soil where the organic and inorganic blend together and the biosphere and atmosphere merge. Freed from his youthful allegiance to Emersonian tran-

EDITOR'S NOTE

scendentalism, Burroughs celebrates the oneness of life found in the earth's natural processes. In contemplating human abuse of this planet and its natural resources, Burroughs takes what is for him a rare critical stance toward modern civilization as he imagines "what a sucked orange the earth will be in the course of a few more centuries."

F.B.

THE GRIST OF THE GODS

ABOUT all we have in mind when we think of the earth is this thin pellicle of soil with which the granite framework of the globe is clothed — a red and brown film of pulverized and oxidized rock, scarcely thicker, relatively, than the paint or enamel which some women put on their cheeks, and which the rains often wash away as a tear washes off the paint and powder. But it is the main thing to us. Out of it we came and unto it we return. "Earth to earth, and dust to dust." The dust becomes warm and animated for a little while, takes on form and color, stalks about recuperating itself from its parent dust underfoot, and then fades and is resolved into the original earth elements. We are built up out of the ground quite as literally as the trees are, but not quite so immediately. The vegetable is between us and the soil, but our dependence is none the less real. "As common as dust" is one of our sayings, but the common, the universal, is always our mainstay in this world. When we see the dust turned into fruit and flowers and grain by that intangible thing called vegetable life, or into the bodies of men and women by the equally mysterious agency of animal life, we think better of it. The trembling gold of the pond-lily's heart, and its petals like carved snow, are no more a transformation of a little black muck and ooze by the chemistry of the sunbeam than our bodies and minds, too, are a transformation of the soil underfoot.

We are rooted to the air through our lungs and to the soil through our stomachs. We are walking trees and floating plants. The soil which in one form we spurn with our feet, and in another take into our mouths and into our blood — what a composite product it is! It is the grist out of which our bread of life is made, the grist which the mills of the gods, the slow patient gods of Erosion, have been so long grinding — grinding probably more millions of years than we have any idea of. The original stuff, the pulverized granite, was probably not very nourishing, but the fruitful hand of time has made it so. It is the kind of grist that improves with the keeping, and the more the meal-worms have worked in it, the better the bread. Indeed, until it has been eaten and digested by our faithful servitors the vegetables, it does not make the loaf that is our staff of life. The more death has gone into it, the more life comes out of it; the more it is a cemetery, the more it becomes a nursery; the more the rocks perish, the more the fields flourish.

This story of the soil appeals to the imagination.

THE GRIST OF THE GODS

To have a bit of earth to plant, to hoe, to delve in, is a rare privilege. If one stops to consider, one cannot turn it with his spade without emotion. We look back with the mind's eye through the vista of geologic time and we see islands and continents of barren, jagged rocks, not a grain of soil anywhere. We look again and behold a world of rounded hills and fertile valleys and plains, depth of soil where before were frowning rocks. The hand of time with its potent fingers of heat, frost, cloud, and air has passed slowly over the scene, and the miracle is done. The rocks turn to herbage, the fetid gases to the breath of flowers. The mountain melts down into a harvest field; volcanic scoria changes into garden mould; where towered a cliff now basks a green slope; where the strata yawned now bubbles a fountain; where the earth trembled, verdure now undulates. Your lawn and your meadow are built up of the ruins of the foreworld. The leanness of granite and gneiss has become the fat of the land. What transformation and promotion! — the decrepitude of the hills becoming the strength of the plains, the decay of the heights resulting in the renewal of the valleys!

Many of our hills are but the stumps of mountains which the hand of time has cut down. Hence we may say that if God made the mountains, time made the hills.

What adds to the wonder of the earth's grist is

that the millstones that did the work and are still doing it are the gentle forces that career above our heads — the sunbeam, the cloud, the air, the frost. The rain's gentle fall, the air's velvet touch, the sun's noiseless rays, the frost's exquisite crystals, these combined are the agents that crush the rocks and pulverize the mountains, and transform continents of sterile granite into a world of fertile soils. It is as if baby fingers did the work of giant powder and dynamite. Give the clouds and the sunbeams time enough, and the Alps and the Andes disappear before them, or are transformed into plains where corn may grow and cattle graze. The snow falls as softly as down and lies almost as lightly, yet the crags crumble beneath it; compacted by gravity, out of it grew the tremendous ice sheet that ground off the mountain summits, that scooped out lakes and valleys, and modeled our northern landscapes as the sculptor his clay image.

Not only are the mills of the gods grinding here, but the great cosmic mill in the sidereal heavens is grinding also, and some of its dust reaches our planet. Cosmic dust is apparently falling on the earth at all times. It is found in the heart of hailstones and in Alpine snows, and helps make up the mud of the ocean floors.

During the unthinkable time of the revolution of the earth around the sun, the amount of cosmic matter that has fallen upon its surface from out the depths of space must be enormous. It certainly must enter largely into the composition of the soil and of the sedimentary rocks. Celestial dirt we may truly call it, star dust, in which we plant our potatoes and grain and out of which Adam was made, and every son of man since Adam — the divine soil in very fact, the garden of the Eternal, contributed to by the heavens above and all the vital forces below, incorruptible, forever purifying itself, clothing the rocky framework of the globe as with flesh and blood, making the earth truly a mother with a teeming fruitful womb, and her hills veritable mammary glands. The iron in the fruit and vegetables we eat, which thence goes into our blood, may, not very long ago, have formed a part of the cosmic dust that drifted for untold ages along the highways of planets and suns.

The soil underfoot, or that we turn with our plow, how it thrills with life or the potencies of life! What a fresh, good odor it exhales when we turn it with our spade or plow in spring! It is good. No wonder children and horses like to eat it!

How inert and dead it looks, yet what silent, potent fermentations are going on there — millions and trillions of minute organisms ready to further your scheme of agriculture or horticulture. Plant your wheat or your corn in it, and behold the miracle of a birth of a plant or a tree. How it pushes up, fed and stimulated by the soil, through the agency of heat and moisture! It makes visible to the eye the life that is latent or held in suspense there in the cool, impassive ground. The acorn, the chestnut, the maple keys, have but to lie on the surface of the moist earth to feel its power and send down rootlets to meet it.

From one point of view, what a ruin the globe is! — worn and crumbled and effaced beyond recognition, had we known it in its youth. Where once towered mountains are now only their stumps — low, fertile hills or plains. Shake down your great city with its skyscrapers till most of its buildings are heaps of ruins with grass and herbage growing upon them, and you have a hint of what has happened to the earth.

Again, one cannot but reflect what a sucked orange the earth will be in the course of a few more centuries. Our civilization is terribly expensive to all its natural resources; one hundred years of modern life doubtless exhausts its stores more than a millennium of the life of antiquity. Its coal and oil will be about used up, all its mineral wealth greatly depleted, the fertility of its soil will have been washed into the sea through the drainage of its cities, its wild game will be nearly extinct, its primitive forests gone, and soon how nearly bankrupt the planet will be!

There is no better illustration of the way decay

and death play into the hands of life than the soil underfoot. The earth dies daily and has done so through countless ages. But life and youth spring forever from its decay; indeed, could not spring at all till the decay began. All the soil was once rock, perhaps many times rock, as the water that flows by may have been many times ice.

The soft, slow, aerial forces, how long and patiently they have worked! Oxygen has played its part in the way of oxidation and dioxidation of minerals. Carbon or carbonic acid has played its part, hydrogen has played its. Even granite yields slowly but surely to the action of rain-water. The sun is of course the great dynamo that runs the earth machinery and, through moisture and the air currents, reduces the rocks to soil. Without solar heat we should have no rain, and without rain we should have no soil. The decay of a mountain makes a hill of fertile fields. The soil, as we know it, is the product of three great processes — mechanical, chemical, and vital — which have been going on for untold ages. The mechanical we see in the friction of winds and waves and the grinding of glaciers, and in the destructive effects of heat and cold upon the rocks; the chemical in the solvent power of rain-water and of water charged with various acids and gases. The soil is rarely the color of the underlying rock from which it came, by reason of the action of the various gases of the atmosphere. Iron is black, but when turned into rust by the oxygen of the air, it is red.

The vital processes that have contributed to the soil we see going on about us in the decay of animal and vegetable matter. It is this process that gives the humus to the soil, in fact, almost humanizes it, making it tender and full of sentiment and memories, as it were, so that it responds more quickly to our needs and to our culture. The elements of the soil remember all those forms of animal and vegetable life of which they once made a part, and they take them on again the more readily. Hence the quick action of wood ashes upon vegetable life. Iron and lime and phosphorus that have once been taken up by growing plants and trees seem to have acquired new properties, and are the more readily taken up again.

The soil, like mankind, profits by experience, and grows deep and mellow with age. Turn up the cruder subsoil to the sun and air and to vegetable life, and after a time its character is changed; it becomes more gentle and kindly and more fertile.

All things are alike or under the same laws — the rocks, the soil, the soul of man, the trees in the forest, the stars in the sky. We have fertility, depth, geniality, in the ground underfoot, on the same terms upon which we have these things in human life and character.

We hardly realize how life itself has stored up

life in the soil, how the organic has wedded and blended with the inorganic in the ground we walk upon. Many if not all of the sedimentary rocks that were laid down in the abysms of the old ocean, out of which our soil has been produced, and that are being laid down now, out of which future soils will be produced, were and are largely of organic origin, the leavings of untold myriads of minute marine animals that lived millions of years ago. Our limestone rocks, thousands of feet thick in places, the decomposition of which furnishes some of our most fertile soils, are mainly of plant and animal origin. The chalk hills of England, so smooth and plump, so domestic and mutton-suggesting, as Huxley says, are the leavings of minute creatures called *Globigerinæ*, that lived and died in the ancient seas in the remote past. Other similar creatures, *Radiolaria* and diatoms, have played an equally important part in contributing the foundation of our soils. Diatom earth is found in places in Virginia forty feet thick. The coral insects have also contributed their share to the soil-making rocks. Our marl-beds, our phosphatic and carbonaceous rocks, are all largely of animal origin. So that much of our soil has lived and died many times, and has been charged more and more during the geologic ages or eternities with the potencies of life.

Indeed, Huxley, after examining the discoveries of the *Challenger* expedition, says there are good grounds for the belief "that all the chief known constituents of the crust of the earth may have formed part of living bodies; that they may be the 'ash' of protoplasm."

This implies that life first appeared in the sea, and gave rise to untold myriads of minute organisms, that built themselves shells out of the mineral matter held in solution by the water. As these organisms perished, their shells fell to the bottom and formed the sedimentary rocks. In the course of ages these rocks were lifted up above the sea, and their decay and disintegration under the action of the elements formed our soil — our clays, our marls, our green sand — and out of this soil man himself is built up.

I do not wonder that the Creator found the dust of the earth the right stuff to make Adam of. It was half man already. I can easily believe that his spirit was evoked from the same stuff, that it was latent there, and in due time, under the brooding warmth of the creative energy, awoke to life.

If matter is eternal, as science leads us to believe, and creation and recreation a never-ending process, then the present world, with all its myriad forms of the organic and the inorganic, is only one of the infinite number of forms that matter must have assumed in past æons. The whole substance of the globe must have gone to the making of other globes such a number of times as no array of figures could express. Every one of the sixty or more primary elements that make up our own bodies and the solid earth beneath us must have played the same part in the drama of life and death, growth and decay, organic and inorganic, that it is playing now, and will continue to play through an unending future.

This gross matter seems ever ready to vanish into the transcendental. When the new physics is done with it, what is there left but spirit, or something akin to it? When the physicist has followed matter through all its transformations, its final disguise seems to be electricity. The solid earth is resolvable into electricity, which comes as near to spirit as anything we can find in the universe.

Our senses are too dull and coarse to apprehend the subtle and incessant play of forces about us — the finer play and emanations of matter that go on all about us and through us. From a lighted candle, or gas-jet, or glowing metal shoot corpuscles or electrons, the basic constituents of matter, of inconceivable smallness — a thousand times smaller than an atom of hydrogen — and at the inconceivable speed of 10,000 to 90,000 miles a second. Think how we are bombarded by these bullets as we sit around the lamp or under the gas-jet at night, and are all unconscious of them! We are immersed in a sea of forces and potentialities of which we hardly dream. Of the scale of temperatures, from absolute zero to the heat of the sun, human life knows only a minute fraction. So of the elemental play of forces about us and over us, terrestrial and celestial —too fine for our apprehension on the one hand, and too large on the other —we know but a fraction.

The quivering and the throbbing of the earth under our feet in changes of temperature, the bendings and oscillations of the crust under the tread of the great atmospheric waves, the vital fermentations and oxidations in the soil — are all beyond the reach of our dull senses. We hear the wind in the treetops, but we do not hear the humming of the sap in the trees. We feel the pull of gravity, but we do not feel the medium through which it works. During the solar storms and disturbances all our magnetic and electrical instruments are agitated, but you and I are all unconscious of the agitation.

There are no doubt vibrations from out the depths of space that might reach our ears as sound were they attuned to the ether as the eye is when it receives a ray of light. We might hear the rush of the planets along their orbits, we might hear the explosions and uprushes in the sun; we might hear the wild whirl and dance of the nebulæ, where suns and systems are being formed; we might hear the "wreck of matter and the crush of worlds" that evidently takes place now and then in the abysms of space, because all these things must send through the ether impulses and tremblings that reach our planet. But if we felt or heard or saw or were conscious of all that was going on in the universe, what a state of agitation we should be in! Our scale of apprehension is wisely limited, mainly to things that concern our well-being.

But let not care and humdrum deaden us to the wonders and the mysteries amid which we live, nor to the splendors and the glories. We need not translate ourselves in imagination to some other sphere or state of being to find the marvelous, the divine, the transcendent; we need not postpone our day of wonder and appreciation to some future time and condition. The true inwardness of this gross visible world, hanging like an apple on the bough of the great cosmic tree, and swelling with all the juices and potencies of life, transcends anything we have dreamed of super-terrestrial abodes. It is because of these things, because of the vitality, spirituality, oneness, and immanence of the universe as revealed by science, its condition of transcending time and space, without youth and without age, neither beginning nor ending, neither material nor spiritual, but forever passing from one into the other, that I was early and deeply impressed by Walt Whitman's lines: —

"There was never any more inception than there is now,
Nor any more youth or age than there is now;
And will never be any more perfection than there is now,
Nor any more heaven or hell than there is now."

And I may add, nor any more creation than there is now, nor any more miracles, or glories, or wonders, or immortality, or judgment days, than there are now. And we shall never be nearer God and spiritual and transcendent things than we are now. The babe in its mother's womb is not nearer its mother than we are to the invisible sustaining and mothering powers of the universe, and to its spiritual entities, every moment of our lives.

The doors and windows of the universe are all open; the screens are all transparent. We are not barred or shut off; there is nothing foreign or unlike; we find our own in the stars as in the ground underfoot; this clod may become a man; yon shooting star may help redden his blood.

Whatever is upon the earth is of the earth; it came out of the divine soil, beamed upon by the fructifying heavens, the soul of man not less than his body.

I never see the spring flowers rising from the mould, or the pond-lilies born of the black ooze, that matter does not become transparent and reveal to me the working of the same celestial powers that fashioned the first man from the common dust.

Man's mind is no more a stranger to the earth than is his body. Is not the clod wise? Is not the chemistry underfoot intelligent? Do not the roots of the trees find their way? Do not the birds know their times and seasons? Are not all things about us filled to overflowing with mind-stuff? The cosmic mind is the earth mind, and the earth mind is man's mind, freed but narrowed, with vision but with erring reason, conscious but troubled, and — shall we say? — human but immortal.

UNDER GENIAL SKIES

In the California Redwoods

EDITOR'S NOTE

Burroughs completed "Under Genial Skies" in the spring of 1920 after spending the winter in southern California. Originally entitled "Pacific Notes," the essay demonstrates the curiosity and skepticism Burroughs maintained into his eighties as he closely examines the nesting habits of trapdoor spiders and questions some theories about desert plants found in John C. Van Dyke's classic work, *The Desert* (1902). In both reading and personal observations Burroughs explored new fields during his trip to the west coast. Although he found the Pacific Ocean at times as forbidding and inhospitable as the desert, he also found it powerfully attractive. That winter he read for the fourth time Richard Henry Dana's *Two Years Before the Mast,* and he enjoyed strolling along the beach and digging for clams. His granddaughter, Elizabeth Burroughs Kelley, reports that he once said everyone should spend at least a month by the ocean, and she remembers his visit with her family on the Massachusetts seashore when he had brought no swimsuit and spent the hot July afternoons wading barefoot through the surf in his long

EDITOR'S NOTE

underwear, rolled up to his knees. When "Under Genial Skies" appeared in *Scribner's* in October 1920, Burroughs said, "It is not fine writing, but it is good enough." Although this essay was the last he saw in print, he continued working on others until he entered the hospital the following February. Just before his final illness he recorded his last entry in the journal he had kept for forty-five years: *Life seems worth living again. Drive to the Sierra Madre P.O. today. Now, at seven pm, I hear the patter of rain.* "Under Genial Skies" was posthumously reprinted in *Under the Maples* in 1921.

F.B.

UNDER GENIAL SKIES

1. A SUN-BLESSED LAND

THE two sides of our great sprawling continent, the East and West, differ from each other almost as much as day differs from night. On the coast of southern California the dominant impression made upon one is of a world made up of three elements—sun, sea, and sky. The Pacific stretches away to the horizon like a vast, shining, gently undulating floor. Its waves are longer and come in more languidly than they do upon the Atlantic coast. It justifies its name. The passion and fury of the Eastern seas I got no hint of, even in winter. Its rocks, all that I saw of them, are soft and friable. The languid waves rapidly wear them down. They are non-strenuous rocks, lifted up out of a non-strenuous sea. The mountains that tower four or five thousand feet along the coast are of the same character. They are young, and while they carry their heads very high, they are soft and easily disintegrated compared with the granite of our coast.

As a rule, young mountains always wear the look of age, from their deep lines and jagged and angular character, while the really old mountains wear the look of youth from their comparative smoothness, their unwrinkled appearance, their long, flowing lines. Time has taken the conceit all out of them.

The annual rainfall in the Far West is only about one third of what it is on the eastern side of the continent. And the soil is curiously adapted to the climate. Trees flourish and crops are grown there under arid conditions that would kill every green thing on the Atlantic seaboard. The soil is clay tempered with a little sand, probably less than ten per cent of it by weight is sand. I washed the clay out of a large lump of it and found the sand a curious heterogeneous mixture of small and large, light and dark grains of all possible forms. The soil does not bake as do our clay soils, and keeps moist when ours would almost defy the plough. Under cultivation it works up into a good tillable condition. Its capacity to retain moisture is remarkable, as if it were made for a scant rainfall. As a crop-producing soil, it has virtues which I am at a loss to account for. Root vegetables grown here have a sweetness, and above all, a tenderness, of which we know nothing in the East. Much sunshine in our climate makes root vegetables fibrous and tough.

I more than half believe that the wonderful sweetness of the bird songs here, such as that of the meadowlark, is more or less a matter of climate; the quality of the sunshine seems to have affected their vocal cords. The clear, piercing, shaft-like note of our meadowlark contrasts with that of the Pacific variety as our hard, brilliant blue skies contrast with the softer and tenderer skies of this sun-blessed land.

To have a smooth grassy lawn about your house on the Pacific coast is to have spread out before you at nearly all hours of the day a pretty spectacle of wild-bird life. Warblers, sparrows, thrushes, titlarks, and plovers flutter across it as thick as autumn leaves—not so highly colored, yet showing a pleasing variety of tints, while the black phœbe flits about your porch and arbor vines.

Audubon's warbler is the most numerous, probably ten to one of any other variety of birds. Then the white-crowned sparrows, Gambel's sparrow, the tree sparrow, and one or two other sparrows of which I am not sure are next in number.

Two species of birds from the Far North are usually represented by a solitary specimen of each, namely, the Alaska hermit thrush and the American pipit, or titlark. The thrush is silent, but has its usual trim, alert look. The pipit is the only walker in the group. It walks about like our oven-bird with the same pretty movement of the head and a teetering motion of the hind part of the body.

While in Alaska, in July, 1899, with the Harriman Expedition, I found the nest of the pipit far up on the side of a steep mountain. It was tucked in under a mossy tuft and commanded a view of sea and mountain such as Alaska alone can afford.

But the most conspicuous and interesting of all these lawn birds are the ring-necked plovers, or killdeers. Think of having a half-dozen or more of those wild, shapely creatures, reminiscent of the shore and of the spirit of the tender, glancing April days, running over your lawn but a few yards from you! Their dovelike heads, their long, slender legs, that curious, mechanical jerking up-and-down movement of their bodies, their shrill, disconsolate cries as they take flight, their beautiful and powerful wings and tail, and their mastery of the air— all arrest your attention or challenge your admiration. They bring the distant and the furtive to your very door. All climes and lands wait upon their wings. They fly around the world.

The plovers are the favored among birds. Beauty, speed, and immunity from danger from birds of prey are theirs. Ethereal and aerial creatures! Is that the cry of the sea in the bird's voice? Is that the motion of the waves in its body? Is that the restlessness of the surf in its behavior?

However high and far it may fly, it has to come back to earth as we all do. It comes to our lawn to feed upon earthworms. The other birds are all busy picking up some minute fly or insect that harbors in the grass, but the plover is here for game that harbors in the turf. His methods are like those of the robin searching for grubs or angle-worms. He scrutinizes the turf very carefully as he runs about over it, making frequent drives into it with his bill, but only now and then seizing the prey of which he is in search. When he does so, he shows the same judgment which the robin does under like conditions. He pulls slowly and evenly, so as to make sure of the whole worm, or to compel it to let go its hold upon the soil without breaking. All birds are wise about their food-supplies.

On the beach the wild life that I see is all on wings. There are the tranquil, effortless gliding herring gulls, snow-white beneath and pearl-gray above, displaying an affluence of wing-power restful to look upon—airplanes that put forth their powers so subtly and so silently as to elude both eye and ear. At low tide I see large groups of their white and gray-blue forms seated upon the dark, moss-covered rocks. Fresh water is at a premium on this coast, and the thirsty gulls avail themselves of the makeshift of the drain-pipes from the town, which discharge on the beach.

There are the clumsy-looking but powerful-winged birds, the brown pelicans, usually in a line of five or six, skimming low over the waves, shaping their course to the "hilly sea," often gliding on set wings for a long distance, rising and falling to clear the water—coasting, at it were, on a horizontal surface, and only at intervals beating the air for more power. They are heavy, awkward-looking birds with wings and forms that suggest none of the grace and beauty of the usual shore birds. They do not seem to be formed to cleave the air, or to part the water, but they do both very successfully. When the pelican dives for his prey, he is for the moment transformed into a thunderbolt. He comes down like an arrow of Jove, and smites and parts the water in surperb style. When he recovers himself, he is the same stolid, awkward-looking creature as before.

A bird evidently not far removed from its reptilian ancestors—a bird that is at home under the water and hunts its prey there on the wing—is the black cormorant. There is a colony of several hundred of them on the face of a sea-cliff a short distance above me.

I see, at nearly all hours of the day, the black lines they make above the foaming breakers as they go and come on their foraging expeditions. In diving, they disappear under the water like the loon, and penetrate to as great depths. One does not crave an intimate acquaintance with them, but they are interesting as a part of the multitudinous life of the shore.

III. SILKEN CHAMBERS

The trap-door spider has furnished me with

one of the most interesting bits of natural history I have found on the coast. An obliging sojourner near me from one of the Eastern States had discovered a large plot of uncultivated ground above the beach that abounded in the hidden burrows of these curious animals. One afternoon he volunteered to conduct me to the place.

The ground was scantily covered with low bushy and weedy growths. My guide warned me that the quarry we sought was hard to find. I, indeed, found it so. It not only required an "eye as practiced as a blind man's touch," it required an eye practiced in this particular kind of detective work. My new friend conducted me down into the plot of ground and, stopping on the edge of it, said, "There is a nest within two feet of me." I fell to scrutinizing the ground as closely as I knew how, fairly bearing on with my eyes; I went over the soil inch by inch with my eyes, but to no purpose. There was no mark on the gray and brown earth at my feet that suggested a trap-door, or any other device. I stooped low, but without avail. Then my guide stooped, and with a long needle pried up a semi-circular or almost circular bit of the gray soil nearly the size of a silver quarter of a dollar, which hinged on the straight side of it, and behold—the entrance to the spider's castle! I was not prepared for any thing so novel and artistic—a long silken chamber, about three quarters of an inch in diameter, concealed by a silken trap-door, an inch in its greatest diameter. The under side of the door, a dull white, the color of old ivory, is slightly convex, and its top is a brownish gray to harmonize with its surroundings, and slightly concave. Its edges are beveled so that it fits into the flaring or beveled end of the chamber with the utmost nicety. No joiner could have done it better. A faint semicircular raised line of clay as fine as a hair gave the only clue. The whole effect, when the door was held open, was of a pleasing secret suddenly revealed.

Then we walked about the place, and, knowing exactly what to look for, I gave my eyes another chance, but they were slow to profit by it. My guide detected one after another, and when I failed, he would point them out to me. But presently I caught on, as they say, and began to find them unaided.

We often found the lord of the manor on duty as doorkeeper, and in no mood to see strangers. He held his door down by inserting his fangs in two fine holes near the edge and bracing himself, or, rather, herself (as, of course, it is the female), offered a degree of resistance surprising in an insect. If one persists with a needle, there is often danger of breaking the door. But when one has made a crack wide enough to allow one to see the spider, she lets go her hold and rushes farther down in her burrow.

Occasionally we found one about half the usual size, indicating a young spider, but no other sizes. My guide said they only emerge from their tunnel at night, and proved it by an ingenious mechanical device made of straws attached to the door. When the door was opened, the straws lifted up, but did not fall down when it was closed. Whenever he found the straw still up in the morning he knew the door had been opened in the night.

As they are nocturnal in habits, they doubtless prey upon other insects, such as sow-bugs and crickets, which the night brings forth. Two bright specks upon the top of the head appear to be eyes, but they are so small they probably only serve to enable them to tell night from day. I think these spiders are mainly guided by a marvelously acute tactile sense. They probably feel the slightest vibration in the earth or air, unless they have a sixth sense of which we know nothing.

All their work, the building and repairing of their nests, as well as all their hunting, is done by night. This habit, in connection with their extreme shyness, makes the task of getting at their life-histories a difficult one. The inside of the burrow seems coated with a finer and harder substance than the soil in which they are dug. It is made on the spot, the spider mixing some secretion of her own with the clay, and working it up into a finer product.

The trap-door sooner or later wears out at the hinge, and is then discarded and a new door manufactured. We saw many nests with the old door lying near the entrance. The door is made of several layers of silk and clay, and is a substantial affair.

The spider families all have the gift of genius. Of what ingenious devices and arts are they masters! How wide their range! They spin, they delve, they jump, they fly. They are the original spinners. They have probably been on their job since carboniferous times, many millions of years before man took up the art. And they can spin a thread so fine that science makes the astonishing statement that it would take four millions of them to make a thread the caliber of one of the hairs of our head—a degree of delicacy to which man can never hope to attain.

Trap-doors usually mean surprises and stratagems, secrets and betrayals, and this species of the arachnids is proficient in all these things.

The adobe soil on the Pacific coast is as well fitted to the purposes of this spider as if it had been made for her special use. But, as in all such cases, the soil was not made for her, but she is adapted to it. It is radically unlike any soil on the Atlantic coast —the soil for cañons and the rectangular watercourses, and for the trap-door spider. It is a tough, fine-grained homogeneous soil, and when dry docs

not crumble or disintegrate; the cohesion of particles is such that sun-dried brick are easily made
from it.

This spider is found in New Mexico, Arizona,
California, and Jamaica. It belongs to the family
of *Mygalidae*. It resembles in appearance the tarantula of Europe, described by Fabre, and has many
of the same habits; but its habitation is a much
more ingenious and artistic piece of workmanship
than that of its European relative. The tarantula
has no door to her burrow, but instead she builds
about the entrance a kind of breastwork an inch
high and nearly two inches in diameter, and from
this fortress sallies out upon her prey. She sinks
a deeper shaft than does our spider, but excavates
it in the same way with similar tools, her fangs,
and lines it with silk from her own body.

Our spider is an artist, evidently the master
builder and architect of her kind. Considering her
soft and pussy-like appearance—no visible drills for
such rough work—one wonders how she excavates
a burrow six inches or more deep in this hard
adobe soil of the Pacific coast, and how she removes
the dirt after she has loosened it. But she has
been surprised at her work; her tools are her two
fangs, the same weapons with which she seizes and
dispatches her prey, and the rake or the *chelicerœ*.
To use these delicate instruments in such coarse
work, says Fabre, seems as "illogical as it would

to dig a pit with a surgeon's scalpel." And she carries the soil out in her mandibles, a minute pellet at a time, and drops it here and there at some distance from her nest. Her dooryard is never littered with it. It takes her one hour to dig a hole the size of half an English walnut, and to remove the earth.

One afternoon I cut off the doors from two nests and left them turned over, a few inches away. The next morning I found that the occupants of the nests, under cover of the darkness, had each started the construction of a new door, and had it about half finished. It seemed as if the soil on the hinge side had begun to grow, and had put out a semicircular bit of its surface toward the opposite side of the orifice, each new door copying exactly the color of the ground that surrounded it, one gray from dead vegetable matter, the other a light brick-red. I read somewhere of an experimenter who found a nest on a mossy bit of ground protectively colored in this way. He removed the lid and made the soil bare about. The spider made a new lid and covered it with moss like the old one, and her art had the opposite effect to what it had in the first case. This is typical of the working of the insect mind. It seems to know everything, and yet to know nothing, as we use the term "know."

On the second morning, one of the doors had attained its normal size, but not yet its normal thickness and strength. It was much more artfully concealed than the old one had been. The builder had so completely covered it with small dry twigs about the size of an ordinary pin, and had so woven these into it, standing a few of them on end, that my eye was baffled. I knew to an inch where to look for the door, and yet it seemed to have vanished. By feeling the ground over with a small stick I found a yielding place which proved to be the new unfinished door. Day after day the door grew heavier and stronger. The builder worked at it on the under side, adding new layers of silk. There is always a layer of the soil worked into the door to give it weight and strength.

Spiders, like reptiles, can go months without food. The young, according to Fabre, go seven months without eating. They do not grow, but they are very active; they expend energy without any apparent means of keeping up the supply. How do they do it? They absorb it directly from the sun, Fabre thinks, which means that here is an animal between which and the organic world the vegetable chlorophyl plays no part, but which can take at first-hand, from the sun, the energy of life. If this is true, and it seems to be so, it is most extraordinary.

In view of the sex of the extraordinary spider I have been considering, it is interesting to remember that one difference between the insect world and the world of animal life to which we belong, which Maeterlinck has forgotten to point out, is this:

In the vertebrate world, the male rules; the female plays a secondary part. In the insect world the reverse is true. Here the female is supreme and often eats up the male after she has been fertilized by him. Motherhood is the primary fact, fatherhood the secondary. It is the female mosquito that torments the world. It is the female spider that spins the web and traps the flies. Size, craft, and power go with the female. The female spider eats up the male after he has served her purpose; her caresses mean death. The female scorpion devours the male in the same way. Among our wild bees it is the queen alone that survives the winter and carries on the race. The big noisy blow-flies on the window-pane are females. With the honey bees the males are big and loud, but are without any authority, and are almost as literally destroyed by the female as is the male spider. The queen bee does not eat her mate, but she disembowels him. The work of the hive is done by the neuters. In the vertebrate world it is chiefly among birds of prey that the female is the larger and bolder; the care of the young devolves largely upon her. Yes, there is another exception: Among the fishes, the females are, as a rule, larger than the males; the immense number of eggs which they carry brings this about.

There are always exceptions to this dominance of the female in the insect world. We cannot corner Nature and keep her cornered. She would not be Nature if we could. With the fireflies, it is the male that dominates; the female is a little soft, wingless worm on the ground, always in the larval state.

In the plant world, also, the male as a rule is dominant. Behold the showy catkins of the chestnuts, the butternuts, the hazelnuts, the willows, and other trees. The stamens of most flowers are numerous and conspicuous. Our Indian corn carries its panicle of pollen high above the silken tresses which mother the future ear.

One day I dug up a nest which was occupied by a spider with her brood of young ones. I took up a large block of earth weighing ten pounds or more, and sank it in a box of earth of its own kind. I kept it in the house under observation for a week, hoping that at some hour of day or night the spider would come out. But she made no sign. My ingenious friend arranged the same mechanical contrivance over the door which he had used successfully before. But the latch was never lifted. Madam Spider sulked or bemoaned her fate at the bottom of her den. At the end of a week I broke open the nest and found her alone. She had evidently devoured all her little ones.

I kept two nests with a spider in each in the

house for a week, and in neither case did the occupant ever leave its nest.

Apparently the young spiders begin to dig nests of their own when they are about half-grown. As to where they stay, or how they live up to that time, I have no clue. The young we found in several nests were very small, not more than an eighth of an inch long. Of the size and appearance of the male spider, and where he keeps himself, I could get no clue.

One morning I went with my guide down to the spider territory, and saw him try to entice or force a spider out of her den. The morning previous he had beguiled several of them to come up to the opening by thrusting a straw down the burrow and teasing them with it till in self-defense they seized it with their fangs and hung on to it till he drew them to the surface. But this morning the trick would not work. Not one spider would keep her hold. But with a piece of wire bent at the end in the shape of a hook, he finally lifted one out upon the ground. How bright and clean and untouched she looked! Her limbs and a part of the thorax were as black as jet and shone as if they had just been polished. No lady in her parlor could have been freer from any touch of soil or earth-stain than was she. On the ground, in the strong sunlight, she seemed to be lost. We turned her around and tried to induce her to enter the nest again; but over and over she ran across the open door without heeding it. In the novel situation in which she suddenly found herself, all her wits deserted her, and not till I took her between my thumb and finger and thrust her abdomen into the hole, did she come to herself. The touch of that silk-lined tube caused the proper reaction, and she backed quickly into it and disappeared.

Just what natural enemy the trap-door spider has I do not know. I never saw a nest that had been broken into or in any way disturbed, except those which we had disturbed in our observations.

IV.   THE DESERT NOTE

I often wonder what mood of Nature this world of cacti which we run against in the great Southwest expresses. Certainly something savage and merciless. To stab and stab again suits her humor. How well she tempers her daggers and bayonets! How hard and smooth and sharp they are! How they contrast with the thick, succulent stalks and leaves which bear them! It is a desert mood; heat and drought appear to be the exciting causes. The scarcity of water seems to stimulate Nature to store up water in vegetable tissues, just as it stimulates men to build great dams and reservoirs. These giant cacti are reservoirs of water. But why spines and prickles and cruel bayonets? They certainly cannot be for protection or defense; the grass and other vegetation upon which the grazing animals feed are not armed with spines.

If the cacti were created that grazing animals in the desert might have something to feed upon, as our fathers' way of looking at things might lead us to believe, why was that benevolent plan frustrated by the armor of needles and spines?

Nature reaches her hungry and thirsty creatures this broad, mittened hand like a cruel joke. It smites like a serpent and stings like a scorpion. The strange, many-colored, fascinating desert! Beware! Agonies are one of her garments.

All we can say about it is that Nature has her prickly side which drought and heat aggravate. In the North our thistles and thorns and spines are a milder expression of this mood. The spines on the blackberry-bush tend against its propagation for the same reason. Among our wild gooseberries, there are smooth and prickly varieties, and one succeeds about as well as the other. Apple- and pear-trees in rough or barren places that have a severe struggle for life, often develop sharp, thorny branches. It is a struggle of some kind which begets something like ill-temper in vegetation— heat and drought in the desert, and browsing animals and poor soil in the temperate zones. The devil's club in Alaska is one mass of spines; why, I know not. It must just be original sin. Our raspberries have prickles on their stalks, but the large, purple-flowering variety is smooth-stemmed.

Mr. John C. Van Dyke in his work on the desert expresses the belief that thorns and spines are given to the desert plants for protection; and that if no animal were there that would eat them, they would not have these defenses. But I believe if there had never been a browsing animal in the desert the cacti would have had their thorns just the same.

Nature certainly arms her animal forms against one another. We know the quills of the porcupine are for defense, and that the skunk carries a weapon that its enemies dread, but I do not believe that any plant form is armed against any creature whose proper food it might become. Cacti carry formidable weapons in the shape of spines and thorns, but the desert conditions where they are found, heat and aridity, are no doubt their primary cause. The conditions are fierce and the living forms are fierce.

We cannot be dogmatic about Nature. From our point of view she often seems partial and inconsistent. But I would just as soon think that Nature made the adobe soil in the arid regions that the human dwellers there might have material at hand with which to construct a shelter, as that she gives spines and daggers to any of the vegetable forms to secure their safety. One may confute Mr. Van Dyke out of his own mouth. He says:

UNDER THE MAPLES

Remove the danger which threatened the extinction of a family, and immediately Nature removes the defensive armor. On the desert, for instance, the yucca has a thorn like a point of steel. Follow it from the desert to the high tropical table-lands of Mexico where there is plenty of soil and moisture, plenty of chance for yuccas to thrive, and you will find it turned into a tree and the thorn merely a dull blade-ending. Follow the sahuaro and the pitahaya into the tropics again, and with their cousin, the organ cactus, you will find them growing a soft thorn that would hardly penetrate clothing.

But are they not just as much exposed to browsing animals in the high table-lands as in the desert, if not more so?

Mr. Van Dyke asserts that Nature is more solicitous about the species than about the individual. She is no more solicitous about the one than the other. The same conditions apply to all. But the species are numerous; a dozen units may be devoured while a thousand remain. A general will sacrifice many soldiers to save his army, he will sacrifice one man to save ten, but Nature's ways are entirely different. Both contending armies are hers, and she is equally solicitous about both. She wants the cacti to survive, and she wants the desert animals to survive, and she favors both equally. All she asks of them is that they breed and multiply endlessly. Notwithstanding, according to Van Dyke, Nature has taken such pains to protect her desert plants, he yet confesses that, although it seems almost incredible, it is nevertheless true that "deer and desert cattle will eat the cholla—fruit, stem, and trunk—though it bristle with spines that will draw blood from the human hand at the slightest touch."

This question of spines and thorns in vegetation is a baffling one because Nature's ways are so unlike our ways. Darwin failed utterly in his theory of the origin of species, because he proceeded upon the idea that Nature selects as man selects. You cannot put Nature into a formula.

Behold how every branch and twig of our red thorn bristles with cruel daggers! But if they are designed to keep away bird or beast from eating its fruit, see how that would defeat the tree's own ends! If no creature ate its little red apples and thus scattered its seeds, the fruit would rot on the ground beneath the branches, and the tribe of red thorns would not increase. And increase alone is Nature's end.

It is safe to say, as a general statement, that the animal kingdom is full of design. Every part and organ of our bodies has its purpose which serves the well-being of the whole. I do not recall any character of bird or beast, fish or insect, that does not show purpose, but in the plant world Nature seems to allow herself more freedom, or does not work on so economical a plan. What purpose do the spines on the prickly ash serve? or on the thistles? or on the blackberry, raspberry, gooseberry bushes? or the rose? Our purple-flowering raspberry has no prickles, and thrives as well as any. The spines on the blackberry and raspberry do not save them from browsing cattle, nor their fruit from the birds. In fact, as I have said, the service of the birds is needed to sow their seeds. The devil's club of Alaska is untouchable, it is so encased in a spiny armor; but what purpose the armor serves is a mystery. We know that hard conditions of soil and climate will bring thorns on seedling pear-trees and plum-trees, but we cannot know why.

The yucca or Spanish bayonet and the century-plant, or American aloe (*Agave americana*), are thorny and spiny; they are also very woody and fibrous; yet nothing eats them or could eat them. They are no more edible than cordwood or hemp ropes. This fact alone settles the defense question about spines.

V. SEA-DOGS

THERE is a bit of live natural history out here in the sea in front of me that is new and interesting. A bunch of about a dozen hair seals have their rendezvous in the unstable waves just beyond the breakers, and keep together there week after week. To the naked eye they seem like a group of children sitting there on a hidden bench of rock, undisturbed by the waves that sweep over them. Their heads and shoulders seem to show above the water, and they appear to be having a happy time.

Now and then one may be seen swimming about or lifted up in a wall of green-blue transparent water, or leaping above the wrinkled surface in the exuberance of its animal spirits. I call them children of the sea, until I hear their loud barking, and then I think of them as dogs or hounds of the sea. Occasionally I hear their barking by night when it has a half-muffled, smothered sound.

They are warm-blooded, air-breathing animals, and there seems something incongruous in their being at home there in the cold briny deep— badgers or marmots that burrow in the waves, wolves or coyotes that hunt their prey in the sea.

Their progenitors were once land animals, but Darwinism does not tell us what they were. The whale also was once a land animal, but the testimony of the rocks throws no light upon its antecedents. The origin of any new species is shrouded in the obscurity of whole geological periods, and the short span of human life, or of the whole human history, gives us no adequate vantage-ground from which to solve the problem.

I can easily believe that these hair seals are close akin to the dog. They have five digits; they bolt their food like dogs; their sense of smell is said to be very acute, though how it could serve them in the sea does not appear. The young are born upon the land and enter the water very reluctantly.

This seal is easily tamed. It has the intelligence of the dog and attaches itself to its master as does the dog. Its sense of direction and locality is very acute. This group of seals in front of me, day after day, and week after week, returns to the same spot in the ever-changing waters, without the variation of a single yard, so far as I can see. The locality is purely imaginary. It is a love tryst, and it seems as if some sixth sense must guide them to it. Locality is as unreal in the sea as in the sky, but these few square yards of shifting waters seem as real to these seals as if they were a granite ledge. They keep massed there on the water at that particular point, with their flippers protruding above the surface, as if they were as free from danger as so many picnickers. Yet something attracts them to this particular place. I know of no other spot along the coast for a hundred miles or more where the seals congregate as they do here. What is the secret of it? Evidently it is a question of security from their enemies. At this point the waves break much farther out than usual, which indicates a hidden reef or bench of rocks, and comparatively shallow water. This would prevent their enemies, sharks and killer whales, from stealing up beneath them and pulling them down. I do not hear their barking in the early part of the night, but long before morning their half-muffled baying begins. Old fishermen tell me that they retire for the night to the broad belts of kelp that lie a hundred yards or more out to sea. Doubtless the beds of kelp also afford them some protection from their enemies. The fishermen feel very bitter toward them on account of the fish they devour, and kill them whenever opportunity offers. Often when I lie half asleep in the small hours of the morning, I seem to see these amphibian hounds pursuing their quarry on the unstable hills and mountains of the sea, and giving tongue at short intervals, as did the foxhounds I heard on the Catskills in my youth.

BIBLIOGRAPHICAL NOTE

Information about John Burroughs is available in numerous published and unpublished sources. What follows is a selective list of sources cited or used in the introductory material of this book, along with supplementary material related to Burroughs and the natural history tradition in America.

The most extensive collection of Burroughs papers is in the Manuscript Collection of the Vassar College Library. The collection contains the fifty-three unpublished journals Burroughs kept for the last forty-five years of his life, between 1876 and 1921. The collection also contains photographs, articles, reviews, and over three thousand letters, postcards, and other items of correspondence to and from Burroughs between 1854 and 1921.

The Berg Collection of the New York Public Library contains the early Burroughs notebooks from 1854 to 1876 as well as correspondence that includes approximately three hundred letters from the Burroughs-Benton correspondence between 1862 and 1902.

BIBLIOGRAPHICAL NOTE

Additional libraries and other institutions in the United States containing Burroughs correspondence and manuscript materials include the New York Historical Society, Pierpont Morgan Library, American Museum of Natural History, American Academy of Arts and Letters, Library of Congress, Huntington Library, Hartwick College, Middlebury College, Mount Holyoke College, Bowdoin College, Smith College, Stanford University, Yale University, University of Virginia, University of Texas at Austin, and the University of Pennsylvania. Other repositories include Greenfield Village in Dearborn, Michigan; Jones Public Library in Amherst, Massachusetts; St. John's Seminary in Camarillo, California; and the Farida Wiley Library at the John Burroughs Nature Sanctuary in West Park, New York.

Published biographical information about Burroughs is available in several books, most comprehensively in Clara Barrus, *The Life and Letters of John Burroughs*, two volumes (Boston: Houghton Mifflin, 1925). Other books by Barrus, Burroughs's authorized biographer, include *Our Friend John Burroughs* (Boston: Houghton Mifflin, 1914); *John Burroughs, Boy and Man* (New York: Doubleday, Page, 1920): *The Heart of Burroughs's Journals*, ed. Clara Barrus (Boston: Houghton Mifflin, 1928); and

BIBLIOGRAPHICAL NOTE

Whitman and Burroughs, Comrades (Boston: Houghton Mifflin, 1931).

Contemporary responses to Burroughs and his work include William Dean Howells, "Wake-Robin," *Atlantic* 28 (August 1871); Theodore Dreiser, "John Burroughs in His Mountain Hut," *The New Voice* 16 (August 19, 1899); Elbert Hubbard [Fra Elbertus, pseud.], *Old John Burroughs* (East Aurora, N.Y.: Roycroft Shop, 1901); Henry James, "A Note on John Burroughs," in *Views and Reviews* (1908; Freeport, N.Y.: Books for Libraries Press, 1968); Dallas Lore Sharp, "Fifty Years of John Burroughs," *Atlantic* 106 (November 1910); and R. J. H. De Loach, *Rambles with John Burroughs* (Boston: Gorham, 1912).

Reminiscences of Burroughs appear in several books, including Julian Burroughs, *My Boyhood, With a Conclusion* (New York, 1922); Hamlin Garland, *Roadside Meetings* (New York: Macmillan, 1930) and *Afternoon Neighbors* (New York: Macmillan, 1934); H. A. Haring, ed., *The Slabsides Book of John Burroughs* (Boston: Houghton Mifflin, 1931): Clifton Johnson, *John Burroughs Talks: His Reminiscences and Comments* (Boston: Houghton Mifflin, 1922); and Elizabeth Burroughs Kelley, *John Burroughs' Slabsides* (Rhinebeck, N.Y.: Moran, 1974).

Other books containing information about Burroughs's friendship with Whitman are Gay Wilson Allen, *The Solitary Singer: A Critical Biography of Walt Whitman,* rev. ed. (New York: New York University Press, 1955) and Justin Kaplan, *Walt Whitman: A Life* (New York: Simon and Schuster, 1980). Whitman's own accounts of Burroughs as one of his closest friends and of his visits to Burroughs's home at Riverby are in *Specimen Days* (Philadelphia: David McKay, 1882). An account of Burroughs's role as official historian of the Harriman Expedition is in William H. Goetzmann and Kay Sloan, *Looking Far North: The Harriman Expedition to Alaska, 1899* (New York: Viking, 1982).

Among scholarly studies containing assessments of Burroughs's writings are Norman Foerster, *Nature in American Literature* (New York: Macmillan, 1923); Philip M. Hicks, *The Development of the Natural History Essay in American Literature* (Philadelphia: University of Pennsylvania Press, 1924); Robert Henry Welker, *Birds and Men: American Birds in Science, Art, Literature, and Conservation, 1800–1900* (Cambridge: Harvard University Press, 1955); Hans Huth, *Nature and the American: Three Centuries of Changing Attitudes* (Berkeley: University of California Press, 1957); Howard Mumford Jones, *The Age of Energy: Varieties of American Experience, 1865–1915* (New York: Viking, 1971);

BIBLIOGRAPHICAL NOTE

H. R. Stoneback, "John Burroughs: Regionalist," in *The Literature of the Mid-Hudson Valley*, ed. Alfred H. Marks (New Paltz, N.Y.: State University of New York at New Paltz, 1973); Perry D. Westbrook, *John Burroughs*, Twayne United States Author Series (New York: G. K. Hall, 1974); David Pierce, James Stapleton, and Perry D. Westbrook, in "A Symposium in Honor of John Burroughs," Bard / Hudson Valley Studies (Annandale-on-Hudson, N.Y.: Bard College, 1979); and Neal Steiger, "Whitman and Burroughs: Comrades in Criticism as Demonstrated by the Themes of Calamus" (M.A. thesis, State University of New York at New Paltz, 1983).

Studies and bibliographies of the ecological sciences in their early development include Eugene Cittadino, "Ecology and the Professionalization of Botany in America, 1890–1905," *Studies in the History of Biology* 4 (1980); Frank N. Egerton, "Changing Concepts of the Balance of Nature," *The Quarterly Review of Biology* 48 (1973) and "The History of Ecology: Achievements and Opportunities," *Journal of the History of Biology* 16 (1983); Robert M. McIntosh, "The Background and Some Current Problems of Theoretical Ecology," *Synthese* 43 (1980), and *The Background of Ecology: Concept and Theory* (New York: Cambridge University Press,

BIBLIOGRAPHICAL NOTE

1985); Daniel Simberloff, "A Succession of Paradigms in Ecology," *Synthese* 43 (1980); and Donald Worster, *Nature's Economy: A History of Ecological Ideas* (1977; New York: Cambridge University Press, 1985).

Some popular accounts of American naturalists which include considerations of Burroughs are Robert Elman, *First in the Field: America's Pioneering Naturalists* (New York: Mason / Charter, 1977); Wayne Henley, *Natural History in America* (New York: Quadrangle / New York Times, 1977); and Paul Brooks, *Speaking for Nature: How Literary Naturalists from Henry Thoreau to Rachel Carson Have Shaped America* (Boston: Houghton Mifflin, 1980).

The most recent Burroughs biography is Elizabeth Burroughs Kelley, *John Burroughs: Naturalist, the Story of His Work and Family* (1959; West Park, N.Y.: Riverby Books, 1986). As curator and owner of the largest private collection of Burroughs material, Elizabeth Burroughs Kelley has made her resources available to numerous students of her grandfather's work and it is to her that this edition is dedicated.

BOOKS BY JOHN BURROUGHS

Notes on Walt Whitman as Poet and Person, 1867
Wake-Robin, 1871
Winter Sunshine, 1875
Birds and Poets, 1877
Locusts and Wild Honey, 1879
Pepacton, 1881
Fresh Fields, 1884
Signs and Seasons, 1886
Indoor Studies, 1889
Riverby, 1894
Whitman: A Study, 1896
The Light of Day, 1900
John James Audubon, 1902
Literary Values, 1902
Far and Near, 1904
Ways of Nature, 1905
Bird and Bough, 1906
Camping with President Roosevelt, 1906,
expanded and reissued as
Camping and Tramping with Roosevelt, 1907
Leaf and Tendril, 1908
Time and Change, 1912

The Summit of the Years, 1913
The Breath of Life, 1915
Under the Apple Trees, 1916
Field and Study, 1919
Accepting the Universe, 1920
Under the Maples, 1921
The Last Harvest, 1922